AF576944

COMBUSTION TOXICOLOGY

COMBUSTION TOXICOLOGY

Shayne C. Gad

G. D. Searle & Co

Skokie, Illinois

and

Rosalind C. Anderson

Anderson Laboratories

Dedham, Massachusetts

CRC Press

Boca Raton   Ann Arbor   Boston

Library of Congress Cataloging-in-Publication Data

Gad, Shayne C., 1948-
Combustion toxicology / Shayne C. Gad and Rosalind C. Anderson.
p. cm.
Includes bibliographical references.
ISBN 0-8493-8802-3
1. Combustion gases -- Toxicology. I. Anderson, Rosalind C. II. Title
RA1247.C65G33 1990
615.9′1--dc20 90-10716
CIP

Developer: Telford Press

This book represents information obtained from authentic and highly regarded sources. Reprinted material is quoted with permission, and sources are indicated. A wide variety of references are listed. Every reasonable effort has been made to give reliable data and information, but the authors and the publisher cannot assume responsibility for the validity of all materials or for the consequences of their use.

Direct all inquiries to CRC Press, Inc., 2000 Corporate Blvd., N.W., Boca Raton, Florida, 33431.

Second Printing 1991

International Standard Book Number 0-8493-8802-3

Library of Congress Card Number 90-10716

Printed in the United States

CONTENTS

Chapter 1 INTRODUCTION 1

Analytical Chemistry Approaches 3
Mathematical Modeling 6
Range of Toxicity Concerns 11
References 13

Chapter 2 A SHORT HISTORY OF COMBUSTION TOXICOLOGY 17

References 21

Chapter 3 BASICS OF INHALATION TOXICOLOGY AND HEAT STRESS PHYSIOLOGY 23

General Principles 23
Types of Exposure 23
Basic Steps 24
The Gas Laws 25
Pressure, Temperature and Volume 25
Density 26
Concentration 26
Nonideal Pressure 28
Behavior of the Secondary Phase in a Gas 29
Basics of Respiratory Physiology 29
Oxygen Diffusion and Solubility 33
Gas Volumes and Pressures 34
The Mammalian Respiratory System 36
Structure 36
Lungs 37
Ventilation 38
Transport of Respiratory Gases 40
Hemoglobin 40
Carbon Dioxide Transport 42
Acid-Base Balance 43
Complications of Temperature 46
Influence on Toxicity 46
Physiological or Stress Response 49
Mechanisms of Toxicity 50
Absorption and Distribution 53

Pulmonary Irritation 58
Lexicon 59
References 61

Chapter 4 THE TOXICITY OF SMOKE AND COMBUSTION GASES 63

Postmortem Studies of Fire Victims 64
Toxic Effects of Common Combustion Atmosphere Gases 68
Carbon Monoxide 68
Carbon Monoxide and Sensory Irritants 69
Hydrogen Cyanide 70
Hydrogen Cyanide and Carbon Monoxide 71
Irritant Gases 72
Carbon Dioxide 74
Smoke and Its Toxicity 77
References 78

Chapter 5 COMBUSTION TOXICITY TESTING 81

Terminology and Study Design Factors 81
Basic Toxicology 81
Combustion Terminology 83
Study System Components 84
Combustion Devices 84
The Tube Furnace 84
The Crucible (Cup) Furnace 85
Radiant Heat Devices 85
Exposure Chambers 86
Toxicity Measurements 87
Lethality and LC_{50} Values 87
Incapacitation 90
Sensory Irritation 93
Other Toxicity Indices 94
Analytical Methods 94
Methods for Evaluating the Biological Hazards of Combustion Products 96
Objective of, and Requirements for, Evaluation Methods 96
Screening *Versus* Evaluating 96
Approaches to Evaluating the Toxicity of Combustion Atmospheres 98

Applicable Concepts 99
Concentration-Time Product 99
Haber's Rule 99
Methodology 100
Recommended Test Method 100
Toxicity Measurements 101
Existing Test Systems 103
National Bureau of Standards Method 103
Carnegie Mellon Institute of Research System 109
United States Testing System 112
University of Pittsburg System 112
Federal Aviation Agency System 116
DIN Method 119
Combustion Apparatus 119
Exposure Chamber 119
Biological End Points 120
Atmosphere Analysis 120
Southwest Research Institute (SWRI) system 121
Utah Flamability Research Center System 121
University of Michigan System 122
University of San Francisco System 122
Summary 124
References 124

Chapter 6 REGULATIONS CONCERNING SMOKE TOXICITY 129

Recent Concerns 129
Formal Provisions for Control of Smoke Toxicity 130
Other Regulations Concerning Toxicity of Smoke 131
New York State 132
Test And Test Laboratories 132
Applicability 133
Passing Standards 134
Registration Requirements 134
New York State Fire Prevention Code 9NYCRR 1120, Article 15 134
Registration Format 134
Data Base 135
Public Access 136

New York City 136
Impacted Products 136
Test and Test Laboratories 136
Passing Standards 137
Applicability 138
The Philosophy of Regulation 138
References 140

Chapter 7 A REVIEW OF THE COMBUSTION TOXICITY OF POLYMERS BY CLASS 147

Results To Date (by Class of Polymer) 150
Reference Materials 150
ABS (Acrylonitrile - Butadiene - Styrene) 153
Acrylics 153
Alkyds (Polyesters) 155
Cellulosics 155
Fluoropolymers 156
Polyamides and Polyimides 159
Phenolics 163
Polycarbonates 164
Polyethylenes 164
Polypropylenes 164
Polystyrenes 165
Silicone Polymers 166
Urethanes 166
Vinyls 169
Miscellaneous Minor Polymer Families 170
Latex 171
Polychloroprene (PCP/neoprene) 171
Polyacrylonitrile 171
Polyether Sulfone 171
Polyphosphazene 172
Polyphenylene Oxide 172
References 172

Chapter 8 INFLUENCE OF FLAME RETARDANTS AND OTHER ADDITIVES ON THE TOXICITY OF COMBUSTION PRODUCTS 177

Introduction 177

Classes of Additives 178
Fillers 178
Plasticizers 179
Colorants 179
Stabilizers 179
Processing Aids 180
Curing Agents 180
Antistatic Agents 180
Stain Resistant Agents 180
Classes of Flame/Fire Retardants 181
Antimony Compounds 181
Chlorine Compounds 182
Bromine Compounds 183
Phosphorus Compounds 184
Boron Compounds 184
Aluminum Compounds 185
Molybdenum Compounds 186
Reactive Fire Retardants 186
Summary 187
Toxicity of Flame Retardant Classes 187
Bromine Containing 187
Chlorine Containing 188
Phosphorus Containing 188
Alumina Hydrates 190
Antimony Trioxide 190
References 190

Chapter 9 CURRENT STATUS, ISSUES AND DIRECTIONS IN COMBUSTION TOXICOLOGY 193

Current Status 193
Issues and Directions 196
References 201

Chapter 1

INTRODUCTION

Shayne C. Gad

Synthetic polymers, because of their chemical composition and the chemistry of additives included in them, have the potential, when burned, to release smoke and gases which are qualitatively different from those commonly produced by such natural polymers as wood, cotton and wool (Cullis *et al.*, 1981; Landrock, 1983). Also, the rates of thermal decomposition (and production of decomposition products) of such synthetic materials in a fire may be greater than those of natural materials.

Yet, since the Second World War synthetic polymers have become increasingly used in modern society. Buildings, home furnishings, aircraft interiors, and automobiles all readily reflect this trend. All synthetic polymers may be the site of, or contribute to, a fire. Thus, evaluating the health hazards of the decomposition products formed when plastics are heated (though a relatively new field) is very important. Though early work on simple pyrolysis product toxicity is as much as 35 years old, the oldest true references to combustion toxicology go back only to the early 1950's (Zapp, 1951). Yet today it has become a high priority area for testing within toxicology, which is itself a high priority field. It has had its own annual conferences, a multitude of study committees (both in industry and government), is the subject of various foreign, national, state and municipal regulations, and stands on the verge of becoming the subject of federal regulation. Why?

There are currently some 5,575,000 unintentional fires annually in the U.S. alone, of which only 10% require action by fire departments. Currently, more than 6,000 people die each year in these fires and, as will be demonstrated later, there is good reason to believe that some 80% of these deaths have smoke inhalation (and not the flames themselves) as the primary cause of death (Alexeff and Packham, 1984; Alarie, 1985). We also have a good deal of data to suggest that the proportion dying due to the product inhalation is increasing annually.

It should be remembered, however, that annual fatality rates from fires have actually been going down. In 1933 about 10,000 people lost their lives because of fires. It was observed and commented upon that many of these victims were not burned but succumbed to the effects of 'smoke' and gases but when deaths from this source were reported, it is notable that almost never was it determined if poisonous gas or gases caused the fatality (Ferguson, 1933).

Recent large fires with multiple toxic inhalant-caused deaths (such as the MGM Grand, White Plains Stoffers, Houston Westchase Hilton and Sao Paulo Brazil high-rise fires) have renewed public concern, but should be recognized as not being unique to the last ten years. Four earlier large fires with similar patterns come to mind;

- Newport, Kentucky night club (1978),
- Maury County, Tennessee jail (1977),
- Coconut Grove night club (1942),
- Cleveland Clinic (1929)

Additionally, fire-resultant deaths have occurred in 18 post-crash commercial aircraft since 1965.

As our society (not just in this country but throughout the world) has progressed, we have turned increasingly to synthetic polymers for use in clothes, furnishings, appliances, wall coverings, furniture, carpets, and homes. Not only are these materials less expensive, but in many cases they are more durable, lighter and stronger than the natural products that were used in earlier days and, especially with the addition of fire and flame retardants, their fire performance in buildings is superior. The problem is that when they burn, they can generate gas phase decomposition products which are considerably more toxic than the old natural products (primarily wood). This concern is widespread - international, in fact, as demonstrated by the multitude of reviews of the problem (Autian; 1980; Bott *et al.*, 1969; Einhorn, 1975; Forestier, 1975; Hilado, 1978; Jouany and Roal, 1975; Raftery, 1974; Reinke and Reinhardt, 1973; Repolh *et al.*, 1966; Rumberg, 1971; Sand and Hofmann, 1977; Terril *et al.*, 1977; Vasilev and Iliehkin, 1975; Woolley, 1973) from around the world.

Wood itself generates a multitude of decomposition products when burned. When the combustion is incomplete, the major product in terms of toxicity is carbon monoxide (CO). CO acts in biological systems by competing with oxygen for binding with hemoglobin. When so bound, carboxyhemoglobin (COHb) is formed. As CO has a greater affinity than oxygen for hemoglobin, it competes favorably with and is difficult for oxygen to displace.

Carbon monoxide is the most prevalent toxic component of smoke, and is responsible for the majority of deaths attributed to inhalation in fires or to asphyxia. Other highly toxic gases and aerosols are also produced, however, depending on the composition of the material and the combustion conditions.

The body depends on hemoglobin (in the form of a complex with oxygen called oxyhemoglobin) to transport oxygen to its tissues: if too much hemoglobin (50% or more) is tied up as carboxyhemoglobin then the results, in progressive order, are drowsiness, unconsciousness, and finally death.

The major concern, however, is that the synthetic polymers will either generate a greater amount of carbon monoxide (than wood) under fire conditions or that they will generate quantitatively worse toxic materials (hereafter called for the purposes of this book, toxicants) than wood. Concern initially centered largely on the generation of hydrogen cyanide or of isocyanates (Napier and Wong, 1972), particularly by the various nitrogen containing polymers. Cyanide is of such great concern because it causes death (by a mechanism called histotoxic hypoxia, which involves interference with the cytochrome system in the cells, therefore starving the body of oxygen on a molecular level) at very low levels (1 microgram per mL being a lethal level in the blood).

There now exists a fair body of information both on which gases people are exposed to in fires and what causes kill people in fires. The results of nine of these studies are detailed in Table 1-1.

The picture from all of these reports is not clear especially when the results of experimental large scale fires conducted on purely plastic environments (such as the study of Boudene *et al.* (1978) on a primarily polyvinylchloride - PVC - fire) show that HCl can be a significant threat on its own. What is clear is that (1) decomposition products are a threat; (2) at this point carbon monoxide is the chief life threatening toxicant encountered in fires; and (3) the threat is not, however, limited to carbon monoxide, cyanide, and HCl.

ANALYTICAL CHEMISTRY APPROACHES

There are approaches to evaluating the potential biological hazards of fire gases other than testing with laboratory animals. Most important among these are the analytical chemistry approach and the mathematical modeling approach. Both of these have been actively pursued for more than twenty years in various attempts to derive broad general answers to the problems at hand. The mathematical modeling approach depends on a number of factors or fields of research external to itself, but it is essential to have adequate analytical data

Table 1-1. Human Exposure and Autopsy Data From Fires

Subject Population	Authors	Findings
New York City Fire Fatalities	Zikria; Terrill *et al.*	83% of all casualties resulted from toxicant inhalation. 70% of the deaths resulting from toxicant inhalation were due to carbon monoxide.
State of Maryland Fatalities	Terrill *et al.*, 1978	Of those dying within six hours, 50% died from CO inhalation; HCN and HCl did not appear to be the causes of death in a significant number of cases. In males 40 to 60 years of age, there was a high correlation between death and blood alcohol levels.
Boston City Fire Fighters	Burgess *et al.*; Gold *et al.*	Analysis of multiple samples of the atmospheres firefighters were exposed to showed CO to be an acute hazard and oxygen depletion not to be significant. HCl was present in only 8 of 90 samples, while HCN (at low concentration) was present in half the samples. CO_2 and NO_2 were not present in significant amounts.
Glasgow, Scotland Fire Fatalities	Harland and Woolley, 1979	Blood samples from fatalities identified more than 60 volatile neutral organics. Proprionitrile was identified as being present in measurable quantities in many victims.
Maury County, Tennessee Jailhouse Fire	Birkey *et al.*, 1980	High levels of both HCN and CO were present in victims, but in no case was HCN present in a concentration high enough to be lethal on its own.
English Fire Fatalities (1957-1974)	Bowes, 1974	The proportion of fire casualties due to smoke and toxic gas inhalation had tripled over this 17 year period, and now represents more than half of all fire casualties.
Munich Fire Fatalities	von Meyer *et al.*, 1979	Studied cyanide concentrations in blood of fire victims. The importance of HCN has increased over the years.
Cleveland, City Flash Fire Fatalities	Hirsch *et al.*, 1977	From a small pool of flash fire victims, investigators determined that in this special case carboxyhemoglobin was not connected with the deaths.
Houston Westchase Hilton Fire Fatalities	NFPA	Placement and autopsies on twelve fatalities from this 1982 fire indicated multiple contributing factors in deaths.

to be able to construct a model and the analytical approach has turned out to be enormously complicated (Rasbash; Taylor and Scott; Bott *et al.*, 1969).

Chemical analyses serve a number of purposes in both large and laboratory scale combustion tests. The first and most important purpose of these measurements is to allow the determination of whether observed toxic effects are produced by carbon monoxide or by other toxic decomposition products. Because some studies have indicated a discrepancy between quantities of CO and mortality, the NMAB has recommended that CO levels be monitored in both the combustion atmosphere and in the blood of animals. Another function of these analyses is to provide an indication of the reproducibility and validity of the combustion of the material. Because of the inherent limitations of some combustion device designs, the intended combustion of a material, particularly with larger quantities, does not always occur. Measurements of CO and CO_2 generation can provide evidence of an improper combustion condition and the necessity to repeat an experiment. For example, changes in the CO to CO_2 ratio during an experiment may indicate whether or not the desired combustion mode is occurring in a satisfactory manner. Analyses are also needed to insure that excessive oxygen depletion does not occur during the exposure of animals.

Additionally, of course, accurate analytical measurements of both small and large scale fire test atmospheres are essential to allow any general case analysis of toxicity (as opposed to the special case of the material being tested under the conditions of test) or for there to be any modeling.

It has been found that simple alpha-cellulose results in approximately one hundred and seventy-five different organics under a single thermal decomposition condition (Sumi and Tsuchiya) and that the decomposition products of PVC alone vary tremendously in both nature and proportions of composition with small changes in decomposition temperatures and that the problem of adequately generating, sampling, and analyzing gas phase products from a fire is not an easy one (Boettner and Hall, 1973). The common gases generated in a fire are CO, HCN, NO_2, NH_3, HCl, SO_2, other halogenated species, isocyanates, acrolein and formaldehyde. But unless one wants to limit evaluations, as in the concept of a limiting toxicant (*i.e.*, make a judgment purely on the basis of the most toxic gas present, as was once suggested by Petajan *et al.*, 1975 and 1976), one has to come to the conclusion that it is not possible to make a valid assessment purely on the basis of the principal gases evolved, considered one-by-one. The contributions of heat and of O_2 deprivation, recognized for some time by a number of authors (Chisnall, 1978) and by the National Academy of Sciences Committee which studied the problem (NMAB, 1978), are such that they must be "zeroed out" of all test systems or the studies will become enormously complex. Recently, Morihawa and Yanai (1986) evaluated a limited number of materials in a series of room scale fires, looking at oxygen and heat levels and the concentrations of just eleven toxicants. Carbon monoxide alone was not found adequate to explain lethality and the study was considered a major one. When one then considers that several authors (Gad and Smith, 1983; Grand, 1985) have published studies establishing large effects on test results with changes in single variables (such as heating rate) in the combustion situation, the complexity becomes apparent.

MATHEMATICAL MODELING

Recent years have witnessed the return of some individuals to supporting mathematical models as an effective way to deal with combustion toxicity here and now. Before looking at these recent efforts and models, however, at least two aspects of the underpinnings of such models should be reviewed.

Mathematical models in combustion toxicology are a special case of structure activity relationship (SAR) models. Combustion toxicology itself is a special case of the problem of evaluating the toxicity of mixtures. The problems and principles involved with these two general cases (SAR models and mixtures toxicology) apply equally well to this special case.

Structure activity relationship (SAR) methods have become a legitimate and useful part of toxicology over the last fifteen years or so. These methods are composed of various forms of mathematical or statistical models which seek to predict the adverse biological effects of chemicals based on their structure. The prediction may be of either a qualitative (irritant/nonirritant) or quantitative (LD_{50}) nature, with the second group usually being denoted as QSAR (quantitative structure activity relationship) models. It should be obvious at the outset that the basic techniques utilized to construct such models are mathematical modeling and reduction of dimensionality methods, as discussed in Gad and Weil (1988).

Starting with the initial assumption that there is a relationship between structure and biological activity, we can proceed to more readily testable assumptions.

First, the dose of a chemical species is subject to a number of modifying factors (such as membrane selectivity and selective metabolic action) which are each related in some manner to chemical structure. Indeed, absorption, metabolism, pharmacologic activity and excretion are each subject not just to structurally determined actions, but also to (in many cases) stereo-specific differential handlings.

Given these assumptions, actual elucidation of SAR's requires the following:

a. Knowledge of the biological activities of existing structures.
b. Knowledge of structural features which serve to predict activity (also called molecular parameters of interest).
c. One or more models which relate 2 to 1 with some degree of reliability.

There are a number of approaches to using structural and substructural data and correlating these to biological activity. Such approaches are generally classified as regression analysis methods, pattern recognition methods, and miscellaneous other (such as factor analysis, principal components, and probabalistic analysis).

The regression analysis methods which use structural data have been, as we will see when we survey the state of the art in toxicology, the most productive and useful. "Keys" or fragments of structure are assigned weights as predictors of an activity, usually in some form of the Free - Wilson model which was developed at virtually the same time as the generally better known Hansch. According to this method, the molecules of a chemical series are structurally decomposed into a common moiety (or core) that may be substituted in multiple positions. A series of linear equations are constructed.

The favorable aspects of Free - Wilson type models are:

a. Any set of quantitative biological data may be employed as the dependent variable.
b. No independently determined substituent constants are required.
c. The molecules comprising a sample of interest may be structurally dismembered in any desired or convenient manner.
d. Multiple sites of variable substitution are readily handled by the model.

There are also several limitations; a substantial number of compounds with varying substituent combinations is required for a meaningful analysis; the derived substituent contributions given no reasonable basis for extrapolating predictions from the substituent matrix analyzed; and the model will break down if nonlinear dependence on substituent properties is important or if there are interactions between the substituents.

Pattern recognition methods comprise yet another approach to examining structural features and/or chemical properties for underlying patterns that are associated with differing biological effects. Accurate classification of untested molecules is again the primary goal. This is carried out in two stages. First, a set of compounds, designated the training set, is chosen for which the correct classification is known. A set of molecular or property description features is generated for each compound. A suitable classification algorithm is then applied to find some combination and weight of the descriptors that allows perfect classification. Many different statistical and geometric techniques for this purpose have been presented elsewhere (Gad and Weil, 1986). The derived classification function is then applied in the second step to compounds not included in the training set to evaluate test performance in terms of accuracy of prediction. In published work these have generally been other compounds of known classification also. Performance is judged by the percentage of correct predictions. Stability of the classification function is usually tested by repeating the procedure several times with slightly altered, but randomly varied, sets or samples.

The main difficulty with these methods is in "decoding" the QSAR in order to identify particular structural fragments responsible for the expression of a particular activity; and even if identified as "responsible" for activity, far harder questions for the model to answer are whether the structural fragment so identified is "sufficient" for activity, whether it is always "necessary" for activity, and to what extent its expression is modified by its molecular environment. Most pattern recognition methods use as weighting factors either the presence or absence of a particular fragment or feature (coded 1 or 0), or the frequency of occurrence of a feature. They may be made more sophisticated by coding the spatial relationship between features.

None of our past or current test systems in toxicology work well for predicting effects in the case of mixtures (Gad and Chengelis, 1988). We have data on either pure materials or fairly simple mixtures with a fixed ratio of components. While predicting over a limited range for single structures based on current data can generally be performed with reasonable comfort, the interactions between mixture components as both total amounts and relative proportions vary is beyond our current understanding. Especially since many commercially or environmentally encountered mixtures are very complex, containing in some cases hundreds of components. And, frequently, we are faced with a desire to explore mixtures with a range of amounts of the same components, or with alterations (substitutions) in components.

There can be four general categories of interactions between components - additivity (effect A + effect B = A + B), antagonism (effect A and effect B = A-B), synergism (effect A and effect B = AB) or a combined effect which is qualitatively very different from that of the individual components. Such interactions may also be very time dependent.

Our current understanding of human health and the real world effect of toxicants is a reflection of what is called the multiple causation theory of disease. This holds that each input into the status of an individuals physiological and psychological condition (whether that input is a xenobiotic chemical or a "susceptibility factor"), is a component cause of that status.

Pozzani *et al.* (1959) published data on the toxicity of 36 simple (2 component or "binary") mixtures of vapors to rats, and found that in only two of the cases did the results differ by more than 1.96 standard errors of the estimate from the predictions of Finney's (1952) model for additive/joint action. This model calculates the harmonic mean of the LD_{50} of mixture components as:

$$1/\text{predicted } LD_{50} = Pa/LD_{50} \text{ of component A} + Pb/LD_{50} \text{ of component B}$$

Where Pa and Pb are the proportions of components A and B in the mixture. Smyth *et al.* (1969) later found much the same result for the oral LD_{50}'s of twenty seven pairs of chemicals in rats, and, indeed, for many binary (two component) mixtures the system works well (and is not limited to LD_{50}'s). But most mixtures of concern are multicomponent. It has been proposed that some form of multiple logistic regression model be used for such extrapolations. All of this requires knowledge of the components of the toxicant mixture -in our current case, the combustion atmosphere.

Babrauskas *et al.* (1986) have proposed what they call the N-gas model. This model assumes the following: that the toxicity of fire smoke is determined mainly by a small number of gases and that these may act additively, synergistically, or antagonistically.

Deviations from this postulate will occur infrequently and can be detected using a limited number of animal tests.

The steps in implementing and using this approach are then proposed to be:

- Identifying the initial set of gases to be considered.
- Determining the toxicity of these gases, individually and in combination.
- Combusting samples of test materials under conditions appropriate to the fire scenarios of concern, and measuring the concentrations of the N evolved gases, either continuously or as batch samples.
- Computing the toxicity, usually taken as the LC_{50}, the concentration of combustion products needed to cause 50 percent of the test animals to die after a specified exposure time.
- Running one or two animal tests at that smoke concentration, minus a confidence interval, to ascertain whether an additional, important toxicant is contributing.

- Adding a limited number of new gases as they are shown to have relevance to commercial products, and
- Creating and maintaining a publicly available data base of mixed gas toxicities.

The interaction assumption is that of linear additivity, *i.e.* that the effective concentration to reach some endpoint, which may be the LC_{50} or some other measure, should be evaluated as;

$$\frac{1}{c_{eff}} - \sum_{i} \frac{1}{c_i}$$

where C_i is the LC_{50} or other toxic level value for the individual component i. This simple method does not account for any nonlinearities, such as synergisms or antagonisms. In practice, a more general formulation is used, which has been termed the fractional summation approach, but still has the same limitations.

Tsuchiza and Nakaza (1986) have proposed a version of this linear multiple gas interaction model, and have presented some experimental data supporting its validity in the case of mixtures containing up to three component gases. Tsuchiza (1986) has also published an analysis of various sets of data on carbon monoxide-hydrogen cyanide mixtures showing the effects to be linearly additive. He has proposed that, based on this data, the case of synergistic interactions can be disregarded.

The group of researchers at Southwest Research Institute have published a series of five articles (Kaplan and Hartzell, 1984; Hartzell *et al.* 1985a-e) on modeling the toxicity of fire gases. Based on sets of data on two component gas mixtures, they have proposed what they call the fractional effective dose (FED) model. It starts with the assumption that time-to-effect/concentration experimental data for a toxicant can be plotted as concentration *vs.* the reciprocal of time (1/t). A linear plot is obtained, from which the slope (K) and intercept (b) can be used to calculate EC_{50} or LC_{50} values for any exposure time from the equation

$$C - K\frac{1}{t} + b$$

Components can then be simply summed as to exposure time/ concentration product, and a result in effect predicted based on the totals resulting from such a near-linear combination. Again, only two component models of simple gases (CO, HCN and HCl) have been evaluated under nonfire situations. And other investigators have found that it does not perform well.

RANGE OF TOXICITY CONCERNS

The range of concern for possible health effects of the products of thermal decomposition of plastics is very wide - almost as wide as the entire scope of toxicology. The concerns are outlined in a number of reviews (Eckardt, 1973; Barry and Newman, 1976), and in some cases are as broad as the incineration of plastics in dumps being a major contributor to air pollution in general (van Grimbergen *et al.*, 1971; Boettner *et al.*, 1973).

The most significant of these concerns are those which are acute and short term - death and incapacitation. These will be the major focus of the remainder of this book, so we will only briefly discuss them here. The two are related - indeed the real concern with whether a material generated during a fire will incapacitate someone is that an incapacitated person will not be able to escape a fire, and therefore will die in it. Death from combustion products is primarily a concern only if it occurs during the course of exposure or shortly thereafter for a person surviving until this point (say one hour after a fire) would receive medical attention (in the form of life support, if necessary) sufficient to see him or her through a slower-developing crisis. Incapacitation can (as we will discuss later) occur in several ways. Though most of those doing research in the field are concerned with effects which cause a loss of consciousness, there are others with concerns about effects such as sensory irritation (Barrow *et al.*, 1978) which would serve to incapacitate by making a person unable to escape a fire but would not directly alter consciousness.

The second set of concerns center on what one must consider intermediate health effects - intermediate in the sense that they are not immediately life threatening and are not apparent at the time of exposure, but rather develop or appear over a period of days or weeks. This group includes effects on the cardiovascular system (Thomas and O'Flaherty, 1982), alterations in the activity levels of liver enzymes (Thomas *et al.*, 1978), biochemical and specific morphological changes in the lungs (Heimann *et al.*, 1979) and a condition known as polymer fume fever. All of these concerns are based on effects which have been demonstrated to occur, at least in experimental animals under laboratory conditions. Polymer fume fever, a flu-like, seemingly immunologic-based respiratory condition is associated at least primarily with polytetrafluorethylene (PTFE) or polyvinyl chloride (PVC) at only slightly elevated temperatures. Fume fever has been recognized for a number of years, but is believed to be a serious problem in only a few cases (Williams *et al.*, 1974).

The third and final group of concerns center on long term, genetically involved health effects. A number of authors have identified known mutagenic compounds present either in decomposition product streams or in the parent plastics (Crittenden and Long, 1976; Withey, 1977) or upon testing decomposition products in the Ames test system have found it to be mutagenic (Zitting *et al.*, 1978; Parent *et al.*, 1979). The concern with the presence of mutagenic components is primarily that they will be teratogenic or carcinogenic. Withey has, in fact, proposed that they represent definite hazards in these two areas, while Graf has independently proposed that a carcinogenic risk is present.

Intermediate and long term hazards are, however, of a much lower priority at this point than acute lethality or incapacitation. Because of these priorities, the main stream of research has been directed at short term effects. To begin to understand and to evaluate these tests and the results to date, we shall first define some terms and survey the study designs which are currently being utilized.

As should be apparent by this point, combustion toxicology has grown to the point where it is approaching a transformation. To this point, as is the case with much of toxicology, the field has progressed in the manner of a descriptive science - data has been collected which describes what happens under a variety of situations, but not to explain why things happen. This second step (the development of models which explain why things happen and which therefore serve to predict as yet unobserved happenings) is the mechanistic phase of a science. Progress in combustion toxicology stalled on the verge of its mechanistic phase. The later sections of this book seek to point out the questions to be answered before the field can move to the mechanistic phase. The authors hope this volume will help in this process.

The first question which must be addressed is that of the relationship between the large scale (or "real") fire and the various laboratory models which are currently being employed. Measurements of what goes on in the different regions of a real fire and studies of how particular materials behave in a full scale situation have been made. But only a very few attempts have been made to study the biological effects of large scale fire atmospheres, and these have not been published. Some of the laboratory-scale models attempt to address this problem indirectly by incorporating evaluations of several diverse thermolysis conditions (flaming, nonflaming and 440°C for example) into the evaluations. But a true validation of any of the proposed test systems will require a series of parallel experiments (large scale and laboratory scale) which directly compare results of defined loads of test materials. Only this will verify where the laboratory test results are valid predictors and where they are not. The mathematical modeling efforts for fires, which have been a primary objective of much of the large scale fire testing to date, have (unfortunately) progressed only to the stage of accounting for very simple gas mixture cases, and those not in a fire situation. These modeled situations are even less real-life than the laboratory scale tests themselves.

This brings us to the final point to be raised in considering why combustion toxicology has become such a pressing issue. This is the matter of present and (more important) future regulation of polymers based on the real or perceived hazard posed by their combustion gases. Currently, the International Organization for Standardization (ISO) and the State and City of New York regulate what may be used in buildings based on the results of combustion toxicity tests. The French regulate purely on the basis of having allowable limits on the amounts of N and Cl present in the structure of a building (believing that in so doing they are limiting the amounts of HCN and HCl which may be released in a fire). In an effort stretching over six years a committee gathered by the National Bureau of Standards (NBS) has produced a "standard" test and protocol which several agencies (FAA and CPSC, to name only two) are considering for use as the basis for Federal regulation of polymers in certain uses. New York, however, selected yet another test method.

What must be kept firmly in mind are the limits of these models and tests and their results. Although combustion product toxicity is an important consideration, it should not be the only (or even the most important) factor in regulation; and consideration of toxicity should not be limited to a simple question of acute lethality. The use of combustion toxicity data must be tempered with a knowledge of ignition temperatures and of what portions of the fuel load they would comprise in a fire. The remainder of this volume will hopefully serve to aid in these objectives.

REFERENCES

Alarie, Y. (1985). The Toxicity of Smoke from Polymeric Materials During Thermal Decomposition. *Ann. Rev. Pharmacol. Toxicol.* **25**: 35.

Alexeeff, G. and Packham, S. C. (1984). Evaluation of Smoke Toxicity using Concentration-Time Products. *J. Fire Sci.* **2**: 362.

Autian, John (1970). Toxicologic Aspects of Flamability and Combustion of Polymeric materials. *J. Fire Flamability* **1**: 239.

Babrauskas, V., Levin, B. C. and Gann, R. G. (1986). A New Approach to Fire Toxicity Data for Hazard Evaluation. *ASTM Standardization News* **September, 1986**, pp. 28-33.

Barrow, C. S., Alaire, Y. and Stock, M. F. (1978). Sensory Irritation and Incapacitation Evoked by Thermal Decomposition Products of Polymers and Comparisons with Known Sensory Irritants. *Arch. Environ. Health* **33**: 79.

Barry, T. J. and Newman, B. (1976). Some Problems of Synthetic Polymers at Elevated Temperatures. *Fire Technol.* **12**: 186-192.

Birky, M. M., Paabo, M. and Brown, J. E. (1980). Correlation of Autopsy Data and Materials Involved in the Tennessee Jail Fire. *Fire Saf. J.* **2**: 17-22.

Boettner, E. A. and Ball, G. L. (1973) *Combustion Products from the Incineration of Plastics.* NTIS, Springfield, VA.

Bott, B., Firth, J. G., Jones, T. A. (1969) Evolution of Toxic Gases from Heated Plastics. *Brit. Polym. J.* **1**: 203-4.

Boudene, C., Jouany, J. M. and Truhaut, R. (1978) Etude De La Toxicite Des Produits De Combustion et de Pyrolyse de Materiaux Utilises Dans Le Batiment. *Arch. Des. Malad.* **39**: 3-24.

Bowes, P. C. (1974) Smoke and Toxicity Hazards of Plastics in Fire. *Ann. Occup. Hyg.* **17**: 143-157.

Chisnall, B. (1978) Toxic Gases in Fires. *Fire Prev. Sci. Sc. Tech.* **19**: 16-19.

Critterdan, B. D. and Long R. (1976). The mechanisms of Formation of Polynuclear Aromatic Compounds in Combustion Systems. **In**: *Polynuclear Aromatic Hydrocarbons; Chemistry, Metabolism and Carcinogenesis. A Comprehensive Survey.* (R. Frendenthal and P. W. Jones, Eds.), 209-223.

Cullis, C. F. and Hirschler, M. M. (1981). *The Combustion of Organic Polymers.* Oxford: Clarendon, p. 420.

Echardt, R. E. and Hindin, R. (1973) The Health Hazards of Plastics. *J. Occup. Med.* **15**: 808-819.

Einhorn, I. N. (1975) Physiological and Toxicological Aspects of Smoke Produced During the Combustion of Polymeric Materials. *Environ. Health Perspect* **11**: 163-189

Ferguson, G. E. (1933) Fire Gases. *Quarterly of the National Fire Protection Association* **27**: 110.

Forestier, M. (1975) Combustion Products of Synthetic Materials. *Rev. Tech. Feu.* **16**: 24-6.

Gad, S. C. and Chengelis, C. P. (1988). *Acute Toxicology: Principles and Methods.* Telford Press, Caldwell, NJ.

Gad, S. C. and Smith, A. C. (1983). Influence of Heating Rates on the Toxicity of Evolved Combustion Products: Results and a System for Research. *J. Fire Science* **1**: 465-479.

Gad, S. C. and Weil, C. S. (1988) *Statistics and Experimental Design for Toxicologists. Second Edition.* Telford Press, Caldwell, NJ.

Gold, A., Burgess, W. A. and Clougherty, E. V. (1978) Exposure of Firefighters to Toxic Air Contaminants. *Am. Ind. H . Assoc. J.* **39**: 534-539.

Grand, A.F. (1985). Effect of Experimental Conditions on the Evolution of Combustion Products using a modified University of Pittsburgh Toxicity Test Apparatus. *J. Fire Science* **3**: 280-304.

Harland, W. A. and Woolley, W. D. (1979) *Fire Fatality Study-University of Glasgow*, Information Paper IP 18/79, Building Research Establishment, Glasgow, Scotland.

Hartzell, G. E., Packham, S. C., Grand, A. F. and Switzer, W. G. (1985). Modeling of Toxicological Effects of Fire Gases: III. Quantification of Post-Exposure Lethality of Rats from Exposure to HCl Atmospheres. *J. Fire Science* **3**: 195-206.

Hartzell, G. E., Priest, D. N. and Switzer, W. G. (1985). Modeling of Toxicological Effects of Fire Gases: II. Mathematical modeling of Intoxication of Rats by Carbon Monoxide and Hydrogen Cyanide. *J. Fire Science* **3**: 115-128.

Hartzell, G. E., Stacy, H. W., Switzer, W. G., Priest, D. N. and Packham, S. C. (1985). Modeling of Toxicological Effects of Fire Gases: IV. Intoxication of Rats by Carbon Monoxide in the Presence of an Irritant. *J. Fire Science* **3**: 263-279.

Hartzell, G. E., Switzer, W. G. and Priest, D. N. (1985). Modeling of Toxicologic Effects of Fire Gases: V. Mathematical Modeling of Intoxication of Rats by Combined Carbon Monoxide and Hydrogen Cyanide Atmospheres. *J. Fire Science* **3**: 330-342.

Heimann, R., Skornik, W. and Jaeger, R. J. (1979) Biochemical Changes in Guinea Pig Lungs Following Exposure to Plastic Combustion Products, Paraquat, or ANTU. *Toxicol. Appl. Pharmacol.* 48 A58.

Hilado, Carlos J. (1978) Toxicity of Pyrolysis Gases from Materials. *Sampe Q* **9**: 14-15.

Hirsch, C. S., Bost, R. O., Gerber, S. R., Cowan, M. E., Adelson, L. and Sunshine, I. (1977) Carboxyhemoglobin Concentrations in Flash Fire Victims Without Elevated Carboxyhemoglobin. *Amer. J. Clin. Path.* **68**: 317-320.

Jouany, J. and Raoul, R. (1975) Fire Hazard Evaluation of Both Synthetic and Natural Materials. *Soc. Plast. Eng. Tech. Pap* **21**: 114-117.

Kaplan, H. L. and Hartzell, G. E. (1984). Modeling of Toxicological Effects of Fire Gases: I. Incapacitating Effects of Narcotic Fire Gases. *J. Fire Science* **2**: 286-305.

Landrock, A. H. (1983). *Handbook of Plastics Flamability And Combustion Toxicology. Principles, Materials, Test, Safety and Smoke inhalation Effects.* Noyes, Park Ridge, NJ., p. 308.

Morikawa, T. and Yanai, E. (1986). Toxic Gases Evolution from Air-Controlled Fires in a Semi-Full Scale Room. *J. Fire Science* **4**: 299-314.

Napier, D. H. and Wong, T. W. (1972) Toxic Products from the Combustion and Pyrolysis of Polyurethane Foams. *Br. Polym. J.* **4**: 45-52.

National Materials Advisory Board (1978). *Smoke and Toxicity (Combustion Toxicology of Polymers).* Vol. 3. Publication No. NMAB 318-3, Washington, DC, National Academy of Sciences, pp. 1-55.

NFPA (1983). Twelve Die in Fire at Westchase Hilton Hotel, Houston, Texas. *Fire Journal* **January, 1983**, 10-56.

Parent, R. A., Dilley, J. V. and Simmon, V. F. (1979) Mutagenic Activity of Smoke Condensates from the Non-Flaming Combustion of Ten Flexible Polyurethane Foams Using the Salmonella/Microsome Assay. *J. Combus. Toxicol.* **6**: 256-264.

Petajan, J. H., Baldwin, R. C., Rose, R. F., Jeppsen, R. B. (1976) Assessment of the Relative Toxicity of Materials; Concept of a Limiting Toxicant. *ASTM Spec. Tech. Publ. STD* **614**: 285-99.

Petajan, J. H., Voorhees, K. J., Packham, S. C., Baldwin, R. C., Einhorn, I. N., Grunnet, M. L., Dinger, B. G. and Birky, M. M. (1975) Extreme Toxicity from Combustion Products of a Fire-Retarded Polyurethane Foam. *Science* **187**: 742-7.

Pozzani, U. C., Weil, C. S. and Carpenter, C. P. (1959). The Toxicological Basis of Threshold Limit Values. 5. The Experimental Inhalation of Vapor Mixtures by Rats, with Notes Upon the Relationship Between Single Dose Inhalation and Single Dose Oral Data. *Am. Ind. Hygiene Assoc. J.* **20**: 364-369.

Rasbash, D. J. (1967) Smoke and Toxic Products Produced at Fires. *Plast. Inst. Trans. J.* **2**: 55-60.

Raftery, M. M. (1974) Smoke and Toxicity Hazards of Plastics in Fires. *Nehorlavost Plast. Hmot.* 130-139.

Reinke, R. E. and Reinhardt, C. F. (1973) Fires, Toxicity and Plastics. *Mod. Plast.* **50**: 94-5, 97-8.

Reploh, H., Klosterkoetter, W. and Einch-Rosskamp, P. (1966) Toxicity of Carbonization Products of Synthetics. *Arch. Hyg . Bakteriol.* **150**: 393-405.

Rumberg, Eckhard. (1971) Disintegration Products and Smoke Generation from Fire Action on Plastics. *Ver. Deut. Ing.* **113**: 20-4.

Sand, H. E. and Hofmann, H. T. (1977) Evaluation of Health Hazards Posed by Thermal Decomposition Products of Flamable Materials. *Oesterr Kunstst. Z.* **8**: 37-41.

Smyth, H. F., Weil, C. S., West, J. F. and Carpenter, C. P. (1969). An Explanation of Joint Toxic Action: Twenty-Seven Industrial Chemicals Intubated in Rats in All Possible Pairs. *Tox. Appl. Pharmacol.* **14**: 340-347.

Sumi, K. and Tsuchiya, Y. (1978) Evaluating the Toxic Hazard of Fires. *Can. Build Dig.* 197-1 to 197-4.

Taylor, W. and Scott, K. A. C. (1979) Toxic Gases in Plastics Fires. *Plast. Rubber. Inst.* **95**: 259-266.

Terrill, J. B., Montgomery, R. R. and Reinhardt, C. F. (1977) Devising a Screening Test for Toxic Fire Gases. *Fire Technol.* **13**: 95-104.

Terrill, J. B., Montgomery, R. R. and Reinhardt, C. F. (1978) Toxic Gases from Fires. *Science* **200**: 1343-1347.

Thomas, W. C. and O'Flaherty, E. J. (1982). The Cardiotoxicity of Carbon Monoxide as a Component of Polymer Pyrolysis Smokes. *Tox. Appl. Pharmacol.* **63**: 363-372.

Thomas, W. C., O'Flaherty, E. J., Bell, R. H. and Stemmer, K. L. (1978) The Effect of Polymer Pyrolysis Fumes on the Activity of Rat Hepatic Cytochrome C Oxidase. *Toxicol. Appl. Pharmacol* **45**: 98.

Tuschiza, Y. (1986). On the Unproved Synergism of the Inhalation Toxicity of Fire Gas. *J. Fire Science* **40**: 346-354.

Tsuchiza, Y. and Nakaya, I. (1986). Numerical Analysis of Fire Gas Toxicity: Mathematical Predictions and Experimental Results. *J. Fire Science* **4**: 126-134.

Van Guimbergen, M., Reybrouck, G. and van de Voorde, H. (1971) Development of Impurities in the Air by Burning Thermoplastics. *Zentralbl. Bakteriol. Parasitenk Infektionskr. Hyg.* **155**: 123-130.

Vasilev, G. A. and Ilichkin, V. S. (1975) Evaluation of the Toxicity of Volatile Products from the Combustion of Polymer Materials. *Gig. Sanit.* **5**: 87-91.

Von Meyer, L., Drasch, G. and Kauert, G. (1979) Significance of Hydrocyanic Acid Formation During Fires. *Z. Rechtsmed.* **84**: 69-73.

Williams, N., Atkinson, G. W. and Patchesky, A. S. (1974) Polymer Fume Fever; Not So Benign. *J. Occup. Med.* **16**: 519-526.

Withey, J. R. (1977) Mutagenic Carcinogenic, and Teratogenic Hazards Arising from Human Exposure to Plastics Additives. **In**: *Origins of Human Cancer* **Vol. 4** (H. H. Hiatt, J. D. Watson and J. A. Winsten, Eds.), 219-241.

Wooley, W. D. (1973) Toxic Products from Plastics Materials in Fires. *Plast. Polym.* **41**: 280-6.

Zapp, J. A. (1951) *The Toxicity of Fire.* Chemical Corps., Army Chemical Center, Maryland. Medical Division Sp.

Zikria, B. A., Ferrer, J. M. and Floch, H. F. (1972) The Chemical Factors Contributing to Pulmonary Damage in "Smoke Poisoning". *Surgery* **71**: 704-709.

Zitting, A., Pfaffli, P. and Vainio, H. (1978) Effects of Thermal Decomposition Products of Polystyrene on microsomal Enzymes and Glutathione Levels in Mouse Liver. **In**: *Proceedings of the International Symposium on Styrene: Occupational and Toxicological Aspects.* pp. 16-17 (1978).

Chapter 2

A SHORT HISTORY OF COMBUSTION TOXICOLOGY

Shayne C. Gad

The beginning of combustion toxicology has been ascribed to John Zapp, who, in 1951 reported, in *The Toxicology of Fire*, that fire fatalities were often caused by smoke inhalation and thermal damage to the respiratory tract. Although other scientists became interested in the toxicity of smoke during the next twenty years, it was during the early 1970s that research in the area of smoke toxicity rapidly expanded. There were two areas of activity which caused this impetus. The first was the Federal Trade Commission (FTC) complaint. In October, 1972 the FTC formally announced its inquiry into the field of flamability of plastics. As a result of this action, the Society of the Plastics Industry (SPI) formed a Urethane Safety Group to "foster intelligent regulations, meaningful standards, and truthful communications looking toward safety for the consumer." This group, comprised of high-level executives, quickly raised hundreds of thousand of dollars for research and large-scale tests. It began aggressive efforts to discourage practices contributing to fires, and it hired a public relations firm to promote the industry's image. The USG first met in December, 1972.

In May, 1973, the FTC proposed a complaint that "certain plastic products used in the construction and furnishing of buildings and homes may constitute serious fire hazards." The plastics involved were cellular (foamed) polyurethane and all forms of polystyrene and its copolymers. The Commission alleged that SPI (Society of the Plastics Industry), ASTM (American Society of Test Methods), and 26 producers and marketers, either directly or indirectly:

- Knew of the serious fire hazards since at least 1967, but failed to disclose such material facts to users
- Misrepresented that these combustible plastics are "nonburning" and "self-extinguishing"
- Had knowledge that the test standards permitting these false descriptions are invalid for determining how these products will behave in actual fires

- Failed in their duty to undertake adequate and effective precautionary, remedial and corrective action to eliminate or minimize of risks of injury or damage to life and property

The complaint also alleged that invalid testing standards enabled these plastics to be classified as noncombustible in numerous building codes throughout the country. The complaint alleged that these plastics, once ignited, frequently produce greater fire hazards than traditional materials. Specifically, it was claimed that these plastics, compared with the materials being displaced:

- Spread flame more rapidly and increase the likelihood of "flashover" - when a fire in one portion of a room or structure suddenly engulfs the entire room or structure so that most combustibles in it burn simultaneously and intensively
- Generate extreme heat more quickly
- Produce greater amounts of dense smoke
- Release toxic or flamable gases or chemicals more quickly and release more such gases at various stages of the combustion process

The proposed relief provided that, among other things:

- The alleged misrepresentation be halted
- The use of allegedly invalid tests for burning characteristics be discontinued and their unreliability be made known to purchasers, users, or those who enforce codes embodying such tests
- Clear warnings be given in the future as to possible fire hazards
- A program of precautionary, remedial and corrective action be established to minimize hazards associated with existing installations

As might be expected, there was considerable comment on the FTC action. Both SPI and ASTM claimed that the FTC had no reason to name them in the complaint, and pointed out that they were working to upgrade or devise new testing methods and standards even before the FTC complaint.

It is interesting to note that the 1974 FTC consent decree on urethane foam applied only to rigid urethane foam, and not to the flexible foam. Rigid urethane foams represent only one-fourth of the total urethane foam market. Flexible urethane foam, such as used in upholstered furniture and mattresses ("comfort cushioning"), was not covered because it was not being advertised as nonburning or self-extinguishing.

The second impetus behind the intensified interest in, and concern over, the toxicity of smoke was the finding in 1974 of a neurotoxic organophosphate ester in the smoke produced by combustion of a noncommercial polyurethane foam (Petajan *et al.*, 1975). As a consequence of this finding, the Products Research Committee was formed in 1974, with substantial resources provided by industry for support of research in the behavior of cellular plastics in fire. With funding by this committee as well as by government agencies, numerous laboratories became involved in various areas of fire research.

In November, 1974 a Products Research Committee (on the Fire Safety Aspects of Cellular Plastic Products) (PRC) was organized as a result of the Consent Agreement entered into by the FTC and the Respondents. In this action, the Committee was created to coordinate and manage a scientific research program on the combustibility of cellular plastics for a period of five years. This five-year program was financed by a $5,000,000 Program Fund provided by assessment of the respondents. The objectives of the Committee's research program were: (1) determination of the most effective manner for employing cellular plastic products and systems containing such products to minimize fire hazards in the final intended uses; (2) development of tests as the basis for standards involving large-scale tests, as well as methods by which the results of small-scale tests can be correlated, to provide an index of the behavior of cellular plastics in various burning conditions. The Committee, consisting of a chairman, eight trustees and an Executive Director, went out of existence in 1980. At that time, it had an unused surplus of about $250,000 which it turned over to SPI to be used as seed money to complete its unfinished projects (Grass, 1980).

The PRC finally completed its work, and in April, 1980 and issued a progress report. The report is largely based on 26 reports resulting from grants or contracts supported by PRC. The report gives a total of 17 recommendations, some of which call for action by ASTM, for greater speed in the study of deficiencies in existing tests and adoption of new ones proposed by the Committee. It calls on industry to continue active research on unwanted fires and recommends continued work in developing test methods for smoldering and the toxicity of combustion gases.

As a result of the FTC action in the early 70's, there is no question that suppliers of plastics have been much more careful in describing the flamability properties of their products. ASTM and other standards-developing groups have given considerable effort to developing more meaningful tests and have dropped or modified certain tests. ASTM has discontinued ASTM D 1692-74, Test for the Rate or Extent of Burning of Cellular Plastics. This test procedure gave low burning times and extents of burning for relatively flamable materials. For example, 20-30% of ordinary expanded polystyrene (EPS) samples were found to pass the test with a low rating because of the rapid shrinkage of the thermoplastic foam away from the flame resulted in a failure to ignite adequately for continued burning. The metal screen used as a specimen support contributed considerably to the extinguishment of the sample by removing a significant amount of the heat of combustion. This test method was relatively useless because of the mild test conditions and its inability to distinguish less fire retardant or more flamable compositions.

One of the unfortunate consequences of the sudden availability of research funds was the evolution of a plethora of laboratory test methods for assessing the toxicity of smoke produced by the combustion of materials. Some of these methods were developed by scientists with expertise relevant to combustion methodology but who had little or no training in toxicology. The various methods that evolved during this period utilized different combustion devices and conditions, different animal exposure chambers and, in some cases, different indices of toxicity and different species of test animals. Because of these marked differences, data obtained with one laboratory test method generally had little relevance to data obtained with a different method. In addition, some investigators, unfamiliar with fundamental toxicological principles, have merely compiled and published table after table of meaningless data. A few investigators who appreciated the complexity of smoke atmospheres developed test methods to enable a comprehensive evaluation of the toxicity of combustion products and to understand the mechanisms of their toxicity.

During the 1980s, the emphasis in combustion toxicology moved from the development of test methodologies to the promulgation of a standardized toxicity test method. In 1980, the Center for Fire Research of the National Bureau of Standards (NBS) published the NBS test method (Birky *et al.*, 1980). The method was developed by the NBS, with assistance by an *ad hoc* working group composed of members from academia, industry, and government, for the purpose of establishing a standardized procedure for evaluating the toxicity of combustion products. Investigators began to use this method to compare and rank materials according to the toxicity of their combustion products, a use that is vehemently opposed by the plastics industry. In 1982, the NBS published a modification of the method, with a revised statement of its purpose, such that "this test method is primarily intended for research and preliminary screening purposes" (Levin *et al.*, 1982). At the present time, this method and three other methods are being considered by American Society of Testing Methods (ASTM) task groups for recommendation as a standard ASTM test method. Other significant advances in combustion toxicology during this period include the use of primates in studies of the physiological and behavioral effects of fire gases and combustion atmospheres, the development of mathematical models for predicting the toxic effects of major combustion gases in rodents and the development of computer models for hazard analysis/risk assessment of materials in fires.

During the past two years, a number of state legislatures have considered the adoption of toxicity tests for building materials and furnishings in an effort to reduce fire fatalities due to smoke inhalation. At the present time, only New York State has initiated any action requiring testing. In May, 1984, the Secretary of State for New York made the recommendation that toxicity data on smoke produced by polymeric materials, as determined by the Uniform Fire Prevention and Building Code Council, be filed at the Department of State (Shaffer, 1984). Under proposed Article 15 of New York's Uniform Fire Prevention and Building Code (New York, 1985), manufacturers of regulated products are required to submit combustion toxicity data, in addition to data on ignitability, flame spread, and other properties of materials, to the Secretary of State for filing in a publicly accessible data bank. This requirement met with strong opposition by the plastics industry, but has seemingly prevailed over challenges to it. In the event combustion toxicity testing is not mandated, it is doubtful that any significant effort to evaluate and improve the combustion toxicity of materials will be forthcoming. But even worse would be the adoption of a so-called "toxicity screening test," based solely on lethality in rodents, which would not adequately evaluate the toxicity of combustion products. Such a test would result only in additional cost to the consumer without any significant reduction in the incidence of smoke inhalation deaths.

REFERENCES

Birky, M. M., Paabo, M., Levin, B. C., Womble, S. E., and Malek, D. (1980). Development of Recommended Test Method for Toxicological Assessment of Inhalated Combustion Products. *NBSIR 80-2077*. Washington, DC, National Bureau of Standards.

Levin, B. C., Fowell, A. J., Birky, M. M., Paabo, M., Stolte, A., and Malelz, D. (1982). Further Development of a Test Method for the Assessment of the Acute Inhalation Toxicity of Combustion Products. *NBSIR 82-2 32*, Washington, DC, National Bureau of Standards.

NY Article 15, Part 1120. Combustion Toxicity Testing (1985). Uniform Fire Prevention and Building Code Council, Albany, New York: State of New York, Department of State, 12231.

Petajan, J. H., Voorhees, K. J., Packham, S. C., Baldwin, R. C., Einhorn, I. N., Grunnet, M. L., Dinger, B. G., and Birky, M. M. (1975). Extreme Toxicity From Combustion Products of a Fire-retarded Polyurethane Foam. *Science* **187**: 742.

Shaffer, G. S. (1984) *Fire gas toxicity*. Recommendations of the Secretary of State to the Uniform Fire Prevention and Building Code Council. Albany, NY: State of New York, Department of State, 12231.

Zapp, J. A. (1951). *The Toxicology of Fire.* Medical Division Special Report No. 4. U.S. Army Chemical Center, Maryland p. 92.

Chapter 3

BASICS OF INHALATION TOXICOLOGY AND HEAT STRESS PHYSIOLOGY

Shayne C. Gad

Combustion toxicology is a subdiscipline of inhalation toxicology. Inhalation itself, however, is in many senses a special route of administration for toxicology as a whole and for acute toxicology in particular.

Animal inhalation studies are difficult and complex to perform correctly. Generally, only two broad types of acute studies are performed using this route (the acute systemic exposure and the pulmonary sensitization/irritation study), and, typically, such acute toxicology studies are not performed well. The complexity and cost of such studies firmly dictates that they be performed only when there are substantial opportunities for human inhalation exposure. Combustion toxicology, as we will see, is even more complex.

The special case of combustion toxicology is even more complex and difficult, as any model used must accommodate, or allow for, assessing the effects of both chemical and physical agents (*i.e.*, heat/stress).

GENERAL PRINCIPLES

Types of Exposure

All inhalation studies can be classified in two ways - either by the pattern of exposure or by the physical nature of the contaminant. Both of these classifications are important, as they dictate equipment and animal selection and the details of study design. For combustion toxicology, the physical nature of the contaminant is dictated by the nature of the hazard situation.

Pattern, however, refers to how (or how much of) a test animal is exposed to the contaminated atmosphere of interest. Both in the real world and the laboratory, inhalation exposure is accompanied by dermal and oral exposure. How concerned one is with the possible confounding effects of such other route exposures on the evaluation of biological outcome generally dictates the selection of pattern of exposure. Indeed, some have sought to select pattern because of a desire to simplify the experimental case (by avoiding the variable of heat stress), which is a major component of the combustion toxicology problem.

The three categories of exposure patterns are nose only, head only, and whole body. There are also minor routes (lung only and partial lung) which will not be discussed here and see use only in special research settings, but seek to allow precise delivery of doses of a test material directly to the lungs.

In "nose only", which is the least commonly used (particularly in combustion toxicology) the test animal is situated so that only its nasal region (or, for dogs and primates, where a mask is used to administer test compound, only the mouth and nasal region) is exposed to a test atmosphere. This can be achieved by having the animal restrained with only its nose poking through an elastic barrier or with a breathing mask fitted over the nose and mouth region. There is still some small amount of oral exposure in such a system, as animals will swallow any material deposited on the surface of their mouths or "cleared" from the nasal region or lungs back into the trachea.

Head only exposure, which is somewhat more common, allows a little more potential for oral and dermal exposure. In such a system, the animals entire head, "airlocked" off from the rest of its body by some form of elastic barrier or collar, is in a chamber into which a test atmosphere is introduced.

Whole body exposure, in which the animal is in a chamber into which a test atmosphere is introduced, is the least complicated pattern of exposure. As will be seen later, there are a wide variety of chamber designs available such that all common laboratory species can be exposed using this methodology. There will be extensive dermal and oral administration in animals exposed whole body (particularly oral in rodents and rabbits which carefully "preen" themselves after an exposure). For gases and vapors, of course, such considerations have minimal impact.

Basic Steps

A good inhalation exposure can be broken into four major parts or basic steps. One can consider these as four problems which must be solved before an actual study is undertaken. In fact, it is this aspect of inhalation which makes the proper conduct of such studies difficult and expensive. The steps are:

a. Generation of a test atmosphere - a special case problem of its own in combustion toxicology
b. Containment, mixing and movement of test atmosphere and animals (both before and after exposure)
c. Measurement and characterization of what animals have been exposed to, and
d. Clean up and disposal of resulting "wastes" (gaseous, solid and liquid)

Each of the first three of these steps will be discussed in further detail elsewhere in this volume, while the fourth is beyond the scope of the objectives of this book (the reader is referred to Leong, 1981; Nelson, 1971).

First, however, some basic information on the behavior of gases (and of other phases suspended in gases) and respiratory physiology should be reviewed.

The Gas Laws

Gases are generally the easiest materials to handle. Supplied as either tanks of pressurized gas or as a readily volatilized liquid, it is easy to achieve the desirable characteristics of a generation system (discussed later), usually by simply metering the gas into an airstream, ensuring mixing, and feeding the result into an inhalation chamber.

The behavior of gases is generally described by the gas laws, which assume the gas acts as an "ideal" gas. Under laboratory conditions, such predictions are usually so close to the actual behavior of gases that any variation is minor.

Pressure, Temperature and Volume

The relationship between the pressure and volume of a gas at a constant temperature is described by Boyle's law, which states that volume, V, is inversely proportional to the pressure, P:

$$V = K\left(\frac{1}{P}\right) \qquad \text{(A)}$$

where K is a proportionality constant, or

$$P_1V_1 = P_2V_2 \qquad \text{(B)}$$

The relationship between the temperature and volume of a gas at a constant pressure is described by Charles' law, which states that the volume is directly proportional to the absolute temperature, T:

$$V = KT \qquad \text{(C)}$$

A more general equation involving all three variables can be obtained by combining Equations A and C:

$$PV = KT \qquad \text{(D)}$$

or

$$\frac{P_1V_1}{T_1} = \frac{P_2V_2}{T_2} \quad \text{(E)}$$

If K is proportional to the number of moles of gas, n, then

$$PV = nRT \quad \text{(F)}$$

which is the well-known ideal gas law. The most commonly used values of the molar gar constant, R, can be found in handbooks or in Nelson (1971).

Density

The calculation of gas density is important, both for correcting flow rates and for determining how well gases conform to the ideal gas law. If W is the weight and M is the molecular weight, then

$$n = \frac{W}{M} \quad \text{(G)}$$

Equation F then becomes

$$PV = \frac{WRT}{M} \quad \text{(H)}$$

or

$$\frac{W}{V} = \frac{PM}{RT} \quad \text{(I)}$$

where W/V is the ideal gas density. The calculated densities of some of the more common gases can be found in handbooks or in Nelson (1971).

Concentration

The concentrations of gas and vapor mixtures are almost always expressed in units of percent or parts per million of volume, although in some air pollution, industrial hygiene, and animal toxicology work the most convenient unit may be weight per unit volume. For concentrations above 0.1%, the concentration, C, is usually expressed in percent. This can be calculated for ideal gases by

$$C\% = \frac{10^2V_a}{V_a + V_b + \ldots V_n} = \frac{10^2P_a}{P_a + P_b + \ldots P_n} \quad \text{(J)}$$

where $V_{a,b,\ldots n}$ and $P_{a,b,\ldots n}$ are the volumes and partial pressures of components, a, b, ..., n at a constant temperature. For concentrations below 0.1%, the concentration is usually expressed in parts per million by volume (*i.e.*, the number of parts by volume of trace gas in a million parts of gas mixture). This can be calculated for ideal gases by

$$C_{ppm} = \frac{10^6 V_a}{V_D + V_a} = \frac{10^6 P_a}{P_D + P_a} \qquad \text{(K)}$$

where V_D and P_D are the volume and pressure of the diluent gas. The V_a and the P_a terms are usually neglected when dealing with concentrations below 5000 parts per million since their contribution to the total volume is usually insignificant. However, above 5000 parts per million, the contaminant gas becomes more dominant and its volume must be included in the total gas volume.

Concentration can also be calculated when a known weight or volume of a liquid is vaporized in a known volume of a diluent gas. If the liquid has a weight W and a molecular weight M, and if it is evaporated at a pressure P and a temperature T, then the volume of vapor produced is given by

$$V_a = \frac{WRT}{MP} \qquad \text{(L)}$$

If V_a is diluted with V_p, then the resulting concentration on a volume percent basis is determined by the equation

$$C\% = \frac{10^2 V_a}{V_a + V_d} = \frac{10^2 \frac{WRT}{MP}}{WRT + \frac{V_d}{MP}} = \frac{10^2}{\frac{V_d^m}{1} + WRT} \qquad \text{(M)}$$

Similarly, the concentration in parts per million by volume is determined as

$$C_{ppm} = \frac{10^6 V_a}{V_D} = \frac{10^6 \frac{WRT}{MP}}{V_D} \qquad \text{(N)}$$

Usually, W is expressed in terms of the volume of liquid, V_L and density, P_L. Equation N then becomes

$$C_{ppm} = \frac{10^6 V_L P_L RT}{V_D MP} \qquad \text{(O)}$$

At room temperature (25°C) and pressure (760 mm Hg), this reduces to

$$C_{ppm} = \frac{24.5 \times 10^6 V_L P_L}{V_D M} \qquad \text{(P)}$$

Concentration is sometimes expressed in units of weight per unit volume[1] when the concentration of aerosols such as metal fumes, mists, or dusts in air or another supporting gas is being measured. Copper, lead, and iron fumes are prime examples. Such concentration is calculated by

$$C_{W/V} = \frac{W}{V_D} \qquad \text{(Q)}$$

which is related to concentration in parts per million by

$$C_{W/W} = \frac{10^{-6} C_{ppm} MP}{RT} \qquad \text{(R)}$$

and to vapor pressure by

$$C_{W/V} = \frac{P_a MP}{P_D RT} \qquad \text{(S)}$$

Usually, however, P_D closely approximates P. Equation S, therefore, simplifies to

$$C_{W/V} = \frac{P_a M}{PT} \qquad \text{(T)}$$

Nonideal Pressure

Most gases adhere closely to the ideal gas law at room temperature and atmospheric pressure. However, for the greatest accuracy with mixtures, deviations from ideality must be taken into account, especially at high pressures. This is accomplished by comparing the actual gas density, P_A, with the ideal density P_i. and the actual gas volume V_A with the ideal density, P_P, under the same conditions. The actual gas volume, V_A, can then be corrected to the ideal volume, V_i, and can be used to make precise gas mixtures using the equation

$$V_1 = KV_a = \frac{P_a}{P_1} V_a \qquad \text{(U)}$$

At 0°C and 760 Hg, P_i can be found from

$$P_1 + \frac{M}{22.414} \qquad \text{(V)}$$

1 Weight per unit volume is usually given in units of milligrams per cubic meter.

Tables of values for P_A, P_i and $\frac{p_a}{p_1}$ under standard conditions can be found in Nelson (1971).

Behavior of the Secondary Phase in a Gas

For our purposes aerosol and dust behavior can be described similarly. The two most important characteristics of an aerosol or dust are aerodynamic size (Stoke's diameter) and density. Particle shape, in the special case of fibers (a fiber being defined as a particle with a length greater than three times its width), is also important, as is hydroscopic behavior when it is extreme.

The importance of these characteristics arises from the fact that deep lung penetration is generally important to ensure maximum biological effect, and these characteristics help determine how deep penetration will be. The respiratory tract is generally broken down into three regions - nasopharynx (NP), tracheobronchial region (TB), and pulmonary region (or deep lung). There are five general mechanisms for the deposition of particles in the respiratory tract; three of these are illustrated schematically in Figure 3-1.

The five mechanisms of particle deposition in the respiratory tract are presented in Table 3-1 below.

Some important interrelationship between these mechanisms and secondary considerations are presented in Table 3-2 (after Morrow, 1960).

When a particle comes in contact with the surface of the respiratory tract, the moist surface (and surface tension associated with that surface) prevent the particle (or droplet) from rejoining the airstream, and, therefore, makes contact synonymous with deposition. The importance of the four major factors in deposition of particles in the different regions of the respiratory tract is presented in Table 3-3.

The three major regions of the respiratory tract referred to in Table 3-3 are illustrated (as they occur in humans) in Figure 3-1. Figure 3-2 gives a graphic display of the distribution of deposition of particles/droplets of a range of sizes in each of the three regions.

Basics of Respiratory Physiology

It is the need for oxygen in the final steps of cellular oxidation reductions of the cytochrome system that makes it such an essential material for all animals. Carbon dioxide is a gaseous by-product of the oxidation of protein, carbohydrate, and lipids. Oxygen consumption and carbon dioxide production are nearly inseparable activities, and together they constitute animal respiration.

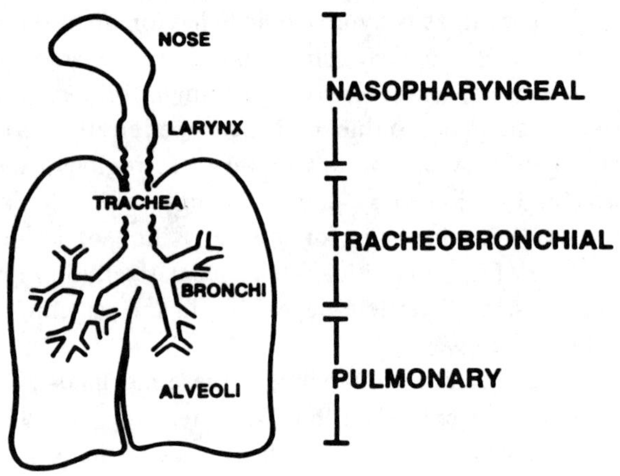

Fig. 3.1

Table 3-1. Mechanism of Respiratory Deposition of Particles

Mechanism	
Direct interception	When a particle physically cannot transit a portion of the respiratory tract Important only for fibers as they pass from one tube to another
Inertial impaction	Particularly important for dense particle Usually occurs near tube bifurcations (*i.e.*, where passages divide)
Sedimentation	Primarily a gravity effect
Diffusion	Particles less than 0.5 microns in diameter and of medium to low density Primarily deep pulmonary
Electrostatic	Special case for unusual particles that carry high charges relative to mass

Table 3-2. Primary Factors Involved in the Respiratory Toxicology of Particulate Matter

Factors	Particles Involved (In Humans)	Related Factors	Deposition Factors	Special Considerations
Sedimentation	> 0.1 μm to <50μm	Particle shape and density. Particulate concentration (aggregation), 'slip', hygroscopicity.	To cause dust deposition in nasal pharynx and tracheolbrochial tree	Highly significant in toxicological studies, probably the most important deposition factor.
Inertia	<50μm	Same as above plus relative velocities of particle and air	Same as above with tendency to promote earlier deposition	Especially significant with soluble or absorbable dusts.
Diffusion	>0.02μm to <0.5μm	Electric charge, hygroscopicity, particle concentration (aggregation), thermal gradients	Dust deposition where large and intimate surfaces involved, such as lung bronchioles and parenchyma.	Most important when dust is insoluble and submicron in size. Also significant when these dusts serve as "carriers".
Respiratory frequency	All sizes	Air velocities, residence times, turbulence, dead space ventilation	Increasing tends decrease deposition (possible very high rates, *e.g.* >20 may increase deposition	Exercise, physical labor and heated environs will have variable effects on subjects but, in general, will tend to increase both these factors together.
Tidal volume	All sizes	Number of particles, alveolar ventilation, residence time	Increasing tends to increase deposition (even at constant minute volume)	

Table 3-3. Relative Importance of Mechanisms

Region	Process			
	Interception	Impaction**	Sedimentation	Diffusion
NP	+++		+	
TB	+	+	+	+
Pulmonary		+	+	+++

** Primarily for fibers

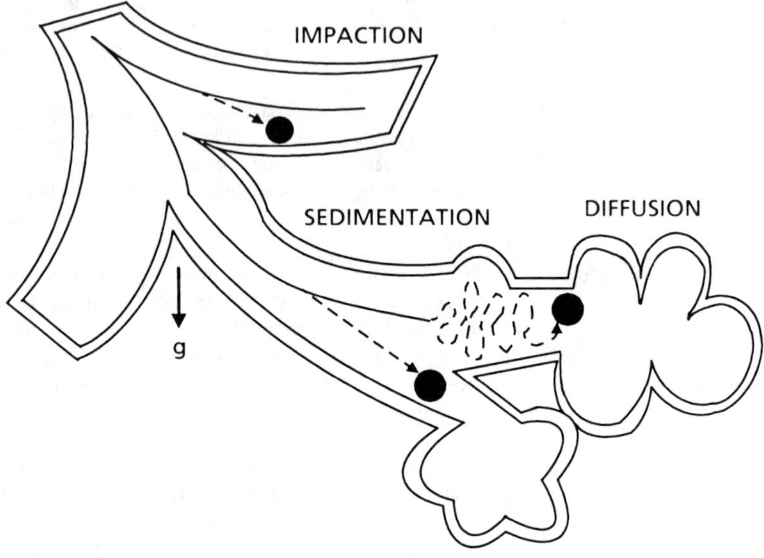

Fig. 3.2

Animal respiration is conveniently divided into three phases:

1. external respiration - the mechanisms by which an animal obtains oxygen from the external environment and the mechanisms by which carbon dioxide is eliminated to the external environment;
2. gas transport - the mechanisms used to distribute oxygen to all of the body cells of an animal after it has been extracted from the external environment by the respiratory organs and also the mechanisms by which carbon dioxide is transported from body cells to sites of elimination;
3. internal respiration - the metabolic reactions of oxidation reduction in which oxygen is consumed and carbon dioxide (and energy) are produced. For our purposes, we are concerned only with the first two of these phases.

To understand animal respiration, we must consider not only the nature of the respiratory organs but also the mechanisms used to control respiration and the nature of respiratory system adaptations to different environments. As with other homeostatic functions, respiration in the animal must be integrated and coordinated with all other regulatory activities. Gas transport mechanisms, in addition to providing a route by which oxygen and carbon dioxide may be transported, also in many animals serve as part of the pH regulating mechanism.

As is true of other homeostatic systems, the respiratory organs and their controls have received the most attention in the vertebrates and especially in mammals. Thus, much of the detail of this discussion will be based on respiration in these groups.

Oxygen Diffusion and Solubility

Oxygen diffuses only very slowly through aqueous media, and diffusion alone is an unsatisfactory mechanism for removing oxygen from the external environment and supplying it to all body cells. Only if the organism is smaller than about 1 mm in diameter is diffusion adequate to supply its oxygen requirements. This does not mean that diffusion of respiratory gases is not an important part of respiratory activity. Diffusion is the basic mechanism by which oxygen or carbon dioxide cross respiratory membranes or move from body fluids into cells; however, diffusion must be aided by other mechanisms that ensure a constant supply of oxygen for the animal and the removal of carbon dioxide. In all larger organisms there is a need for specialized vascular surfaces which extract oxygen from the environment and mechanisms for moving the medium so that the greatest possible oxygen supply is available.

The amount of oxygen required by an animal depends on its size (smaller animals have a higher metabolic rate per unit of body weight and therefore usually require more oxygen on a unit weight basis than do larger animals). However, the activity of an animal also determines its oxygen needs.

Oxygen diffusion is relatively slow in water as compared with its diffusion in air. The diffusion coefficient at 20°C for oxygen in air is about 11; in water the diffusion coefficient is about 0.00003; in a tissue such as muscle the diffusion coefficient is about 0.00001.

The solubility of oxygen in water is much less than its solubility in air. The atmosphere contains about twenty times as much oxygen as can be dissolved in a similar volume of water. This means that, as compared with an air-breather, an aquatic animal must pass a much greater volume of the medium over its respiratory surfaces in order to obtain a given volume of oxygen. In addition, the density and viscosity of aqueous media are greater than those of air, and thus more work must be done to move the aqueous medium in order to obtain oxygen.

The carbon dioxide content of natural water is very low, often zero; and because carbon dioxide is very soluble in water and has a much higher rate of diffusion than oxygen, the concentration gradient of carbon dioxide between the animal and its environment is favorable for the diffusion of carbon dioxide from the animal. Because the terrestrial atmosphere also contains only a very low carbon dioxide concentration, about 0.04 per cent, the elimination of carbon dioxide from an animal is generally not as much of a problem as the obtaining of oxygen. Carbon dioxide in water is converted almost immediately to bicarbonate, and this buffering action serves to maintain a higher pH of aqueous media compared with the pH that would result from the presence of the acidic carbon dioxide itself. The conversion of carbon dioxide to bicarbonate also serves to lower the carbon dioxide concentration of aqueous environments, thus maintaining a favorable concentration gradient for diffusion from the animal.

The oxygen content of water varies according to both temperature and salt concentration. A 2.9 % NaCl solution can contain maximally 40 ml O_2 per liter when equilibrated with oxygen at 0°C and at 1 atmosphere pressure; pure water under these same conditions can hold 50 ml O_2 per liter.

The higher the temperature, the less oxygen (or other gas) a given volume of water can contain. This means that at higher temperatures, the amount of oxygen available to aquatic animals decreases. Concentrations are expressed in terms of the absorption coefficient c, that is, that amount of gas which could dissolve in a given volume of water when the pressure of the gas is 1 atm. The volumes are given in terms of normal temperature (0°C) and pressure (1 atm = 760 mm Hg). The volume of gas that can dissolve decreases in proportion to decreases in its pressure. This is true regardless of the presence of other gases.

Gas Volumes and Pressures

Normal atmospheric pressure at sea level can support a column of mercury 760 mm in height (760 mm Hg pressure = 1 atm). The atmosphere is composed of 78.09 percent nitrogen, 20.95 percent oxygen, 0.93 percent argon, and 0.031 percent carbon dioxide (percentages here are mole % = volume %). The atmosphere contains an average water vapor content with a pressure of 5 mm Hg. On the basis of containing water vapor, the nitrogen content is 79.O percent, oxygen 20.94 percent, and carbon dioxide 0.04 percent. On any basis, the atmosphere is composed almost entirely of nitrogen and oxygen with argon and carbon dioxide making up less than 1 percent of the atmospheric composition.

As discussed earlier, each component in a mixture of gases contributes to the total pressure in direct proportion to its percentage of the composition (Dalton's law, as discussed earlier). The pressure exerted by nitrogen at sea level is: (0.790) (760 mm Hg) =600.55 Hg. Partial pressures are symbolized as P_{N2} where the subscript indicates the gas under discussion. The partial pressure of oxygen in the atmosphere is 159.16 mm Hg, and the partial pressure of carbon dioxide is 0.30 mm Hg. When the total pressure of a gas mixture is 760 mm Hg, the partial pressure of any given gaseous component may be calculated from:

$$P_x = \frac{X}{100}P$$

where P_x is the partial pressure of the gas, P is the total pressure of the gas mixture, and X is the percent volume of the given gas.

The partial pressure of a gas is dependent on the amount of water vapor present in the gas mixture. The higher the temperature, the more water evaporates per unit time, and such water vapor takes up volume within the gas mixture, thus changing the proportion of gases present. Atmospheric gases are usually dried before measurements are made, and their contents are usually expressed in terms of dry air. To calculate the partial pressure of respiratory gases, the water vapor pressure is first subtracted from the total atmospheric pressure.

For example, consider that during inspiration in humans atmospheric air enters the lungs, which are at a temperature of about 37°C, and the atmospheric air becomes saturated with water vapor. To calculate the partial pressure of oxygen, the water vapor pressure at 37°C is subtracted from the atmospheric pressure: 760 - 47 = 713 mm Hg. The P_{CO2} is then equal to (713) (0.2095) = 149.4 mm Hg. On a similar basis, the P_{CO2} is zero, and P_{N2} is 564 mm Hg. Gas volumes, of course, depend on both temperatures and pressure. In order to compare gas volumes, measured volumes are usually converted to normal temperature and pressure (NTP). Normal temperature is taken as 0°C, and normal pressure is 760 mm Hg.

When a gas is dissolved in a liquid, the term gas tension is used to indicate the content of gas within the liquid. Gas tensions are dependent on the partial pressure of the gas above the liquid in the atmosphere and are usually expressed in terms of the gas partial pressure.

THE MAMMALIAN RESPIRATORY SYSTEM

Structure

Because of the relative simplicity of the methods of analysis and its amenability to quantitative analysis, the mammalian respiratory system is probably the best understood physiological system. The structure and function of the human respiratory system will be the one principally described here. It is the best understood and is generally similar to other mammalian respiratory systems.

The lungs serve functions other than that of gas exchange. As already noted, they are involved in temperature regulation and water loss. In addition to providing a site for gaseous exchange, the lungs also serve to warm incoming air and to saturate it with water vapor and they are an important metabolic organ.

Air enters the body through the mouth and nose; the latter possesses mucous membranes that filter foreign particles from the entering air. The air then passes through the pharynx, through the open glottis, through the larynx, and into the trachea - a large tube held open by rings of cartilage that encircle it transversely.

The trachea enters the thoracic cavity and branches into two major air passages, the bronchi, one of which leads to each lung. The bronchi ramify profusely, finally forming smaller tubes, the bronchioles. Except for the smallest terminal bronchioles, all of these tubes have cartilaginous rings. Bronchioles are also equipped with smooth muscle fibers. The terminal bronchioles lead into an expanded portion known as the atrium and from the atrium a number of alveoli open.

The lungs are masses of spongy and elastic tissues and lie in the airtight thoracic cavity. The lungs are enclosed by the visceral pleura, a layer of connective tissue anatomically contiguous with the pericardium. The pleura is another membranous layer lying next to the thoracic cage. Between the two pleura is a thin layer of fluid that lubricates the lungs and membranes during their movements.

Ribs form the side walls of the thoracic cage, and between the ribs are intercostal muscles that can cause movements of the ribs. The base of the thorax is formed from the diaphragm, a sheet of muscle. The diaphragm is cone shaped in the relaxed condition of the diaphragm muscle fibers.

The intercostal muscles are innervated by intercostal nerves derived from the thoracic region of the spinal cord. The diaphragm is innervated by the phrenic nerves derived from a plexus in the cervical region. These muscles and nerves are the main respiratory apparatus. when breathing is heavy, as in exercise, accessory respiratory muscles of the abdominal wall and upper chest are called into play.

Lungs

The lungs are elastic bodies that can return to their normal shape after a deforming force is removed. The inflation of the lungs is accompanied by an increase in potential energy. The conversion of this potential energy into kinetic energy during deflation provides part of the force needed to expel gas. The lungs may be considered as passive elements in the ventilation process, although their elasticity plays a major role in ventilation. It may be noted that the lungs of mammals are inflated by the development of negative pressures, in contrast to the positive pressure systems of frogs and lungfish.

The elasticity of the lungs has been studied by filling them with air or saline solution and measuring the resulting pressure under static conditions. When volume is plotted against pressure, a straight line results whose slope is a measure of the stiffness (compliance), *i.e.*, the change in volume per unit change in pressure per centimeter of water. It is found that the pressure necessary to enlarge the lungs to a given volume is less when the lungs are filled with liquid than when they are filled with air. There is not as much elastic recoil in a liquid-filled lung. Such differences depend on surface tension differences, which in turn are dependent on the nature of the interfaces present. In the liquid-liquid interface condition present when the lungs are filled with a liquid, the surface tension is greatly reduced and depends only on the elasticity of the lungs. When a gas-liquid interface is present, as in the air-filled lung, the compliance includes both the elastic properties of the lungs and the significant surface tension of the interface.

The surface tension is caused by a film of liquid that layers the alveolar surfaces of the lungs. At an interface the attractive forces between molecules are directly downward and sideways more than upward, and this surface force is the surface tension. Laplace showed how the surface tension, T, is related to the pressure, P, and the radius, r in an object such as a soap bubble ($4T=Pr$). If two bubbles have different radii, the pressure in the larger will be less than that of the smaller. And when two such bubbles are connected, the small bubble will empty into the larger. Alveoli have different sizes, and with a uniform surface tension it would be expected that small alveoli would empty into large ones, but this does not normally occur.

The stability of alveoli is produced by the pressure of a surface coating, a surfactant, which causes a nonlinear change in surface tension with surface area. As the lungs are filled with air the alveoli that are most inflated have higher surface tensions than do those that are underdistended. This serves to stabilize alveoli of different sizes.

The surface coating of mammalian alveoli is a complex of protein and dipalmityl lecithin. Whereas surfactants have been most intensively studied in mammals, such substances have also been found in all birds, reptiles, and amphibians studied.

Ventilation

Since the air space in the lungs directly connects with the outside air, the pressure within the lungs (the intrapleural pressure) will be equal to the atmospheric pressure, unless some volume change occurs in the lungs.

Contraction of the inspiratory muscles (the diaphragm and external intercostal muscles) enlarges the thoracic cavity and causes a reduced pressure between the lungs and the thoracic wall (the intrathoracic pressure). This reduction causes the lungs to expand and reduces the intrapleural pressure. The alveoli and bronchioles expand, and air at atmospheric pressure flows through the upper respiratory tract and into the alveoli and bronchioles, where the pressure has been reduced below atmospheric by the expansion.

The intrapulmonic pressure is the force causing movement of air into and out of the lungs and is the pressure gradient between the lung spaces and the atmosphere. When the diaphragm contracts, the sheet of muscle flattens, lowering the floor of the thoracic cavity and thereby increasing its volume. Under these conditions intrapulmonic pressure is reduced, and air flows into the lungs.

The external intercostal muscles cause a rotation of the ribs upward and laterally and a movement of the sternum forward. These actions increase the circumference of the thoracic cavity and thus enlarge its volume also.

To expire air, the inspiratory muscles relax, and the thoracic cavity returns to its normal volume. The elastic recoil of the lungs aid in restoring them to their deflated volume. These forces increase the intrapulmonic pressure and cause air to move from the lungs, through the upper respiratory tract and into the atmosphere.

These are the ventilation movements responsible for bringing fresh air into the lungs and into contact with the alveolar surfaces and for removing the stale air during exhalation.

The volume of air that can be taken into the lungs can be measured by any of the many forms of spirometers. The vital capacity is that volume of air expired by the most forceful expiration after a maximal inspiration. It is the total moveable air in the lungs. The residual volume is the amount of air remaining in the lungs after the most forceful expiration and amounts to about

1.5 liters in humans. The total lung volume is the sum of the vital capacity and the residual volume and amounts to about 6 liters in humans. The tidal volume is the actual amount of air moved at each cycle of inspiration-expiration. It varies according to the needs of the organism and is about 500 ml in humans at rest. The dead space is a volume of about 150 ml in man where air remains in those parts of the respiratory tract, such as the bronchi, where no gas exchange can occur.

All of these ventilation movements are designed to bring fresh air into the alveoli, where oxygen can diffuse down its concentration gradient into the oxygen-depleted venous blood and where carbon dioxide can diffuse from the blood into the air for elimination to the atmosphere.

It is worth noting that a standardized set of symbols has been developed to describe respiratory activity and gas exchange and are used for describing circulatory actions as well. These symbols are, however, beyond the scope of this chapter.

Both the pneumotaxic and apneustic centers are probably innervated by ascending neurons from the medulla. At one time it was thought that stretch-reflexes (the Hering-Breuer reflexes) were responsible for the rhythmic respiratory action. Stretch receptors in the lungs respond to the degree of inflation or stretch of the lungs and send impulses through vagal fibers to the respiratory center of the medulla. These impulses inhibit the inspiratory center, and its output to the inspiratory muscles is stopped. There is a time delay in response to muscle stretch, and inspiration can reach completion before the inspiratory center is inhibited. This is a positive feedback control. it is not considered to create the rhythmicity of breathing because the receptors respond only to relatively extreme degrees of stretch and, additionally, transection of the vagus nerves does not inhibit this rhythmic action.

Other sensory inputs affect breathing rates. Especially important are chemoreceptors and baroreceptors located in the carotid sinus and aortic arch. Impulses from these receptors are carried by the vagus and glossopharyhgeal nerves to the medullary centers. Chemoreceptors of the carotid sinus and aortic arch are sensitive to the P_{O_2} of the blood. They fire continuously at normal blood P_{O_2} levels, and their rate of firing greatly increases when the P_{O_2} is reduced to 50 mm Hg (normal blood P_{O_2} in the arteries is about 100 mm Hg). An increase in the blood P_{CO_2} increases the sensitivity of the chemoreceptors to lowered oxygen levels. Impulses from the chemoreceptors excite the inspiratory center as well as the cardioexcitatory center of the medulla. Baroreceptors are stretch receptors activated by changes in intravascular pressure. A rise in blood pressure increases the frequency of baroreceptor impulses and inhibits respiration rate. However, these receptors are probably more important to circulatory system control than to normal respiratory activity.

The medulla has receptors that are directly sensitive to changes in the P_{CO2} level of the cerebrospinal fluid, which in turn is dependent on the P_{CO2} of the blood. It also appears that the neurons capable of spontaneous discharge and responsible for the basic rhythmic activity of respiration are themselves sensitive to the carbon dioxide levels of blood passing them.

Respiratory control has been analyzed in terms of systems analysis. The respiratory system is, in many respects, part of the excretory system and respiratory control is designed to adequately eliminate CO_2 and also to maintain normal levels of blood CO_2, oxygen, and H^+.

Under conditions of stress accessory muscles of the abdominal wall and upper thorax can be called into activity. Their contractions increase the depth of breathing and thus the amount of air moved into and out of the lungs at each cycle.

TRANSPORT OF RESPIRATORY GASES

Hemoglobin

Once oxygen has diffused from the atmosphere across the respiratory surface and into the blood, it is often transported by means of respiratory pigments. These are required, at least in larger and more active animals, because of the relatively low solubility of oxygen in aqueous solutions. Vertebrate blood plasma can contain maximally about 0.3 ml of O_2 per 100 ml. However, by using transport pigments the concentration of oxygen carried by the blood may amount to 5 to 30 ml O_2 per ml depending on the species.

Oxygen transport pigments are conjugated proteins, that is, proteins complexed with another organic molecule or with one or more metal atoms. Because of the nature of the conjugated group, such transport molecules are colored.

Transport pigments contain metal atoms such as CU^{2+} or Fe^{3+} to which oxygen can reversibly attach. Respiratory pigments are not oxidized by oxygen, rather they are oxygenated, that is, they combine reversibly with molecular oxygen. Respiratory pigments are of value not only because they allow the blood to carry a larger amount of oxygen than would otherwise by possible, but also because they quickly remove oxygen from solution at the respiratory surface, thus maintaining a concentration gradient down which oxygen can diffuse. In addition, it is of value to carry molecular oxygen, O_2, rather than single oxygen atoms.

Further, the oxygenation process allows the pigment to pick up oxygen at sites of high oxygen tension and to release the oxygen at sites of low oxygen tension (body cells requiring oxygen).

In vertebrates the respiratory pigment is hemoglobin. Hemoglobin has a molecular weight of about 68,000 and is composed of two pairs of polypeptide chains. Each chain carries an iron-containing heme group. The hemoglobin molecule is capable of transporting four oxygen molecules. When four oxygens are attached to hemoglobin the resulting complex is termed oxyhemoglobin and is said to be fully saturated with oxygen. The oxygen capacity of blood is defined as the total amount of oxygen that can be taken up by a unit volume of the blood and includes that oxygen transported by respiratory pigments as well as that in solution.

At a given pH the amount of oxygen that a given quantity of hemoglobin can pick up depends on the partial pressure of oxygen. The oxygen pressure is usually given in units of mm Hg, and the loading of hemoglobin is given as the percentage of saturation with oxygen. Such plots are oxygen dissociation curves. At about 40 mm Hg oxygen tension a small reduction in the P_{O2} will allow a relatively large amount of oxygen to be released by the hemoglobin. Because 40 mm Hg is about the P_{O2} of body fluids and tissues, hemoglobin is adapted to release its oxygen at pressures close to those of the cells that need it. At higher P_{O2} levels, which fall in the range of alveolar oxygen concentrations, hemoglobin becomes completely saturated with oxygen.

The term half-saturation pressure is given to the P_{O2} at which hemoglobin (or any respiratory pigment) is only half-saturated with oxygen. A respiratory pigment with a lower half-saturation pressure has a higher affinity for oxygen when compared with a pigment having a higher half-saturation pressure. It is of interest to compare the muscle oxygen-storage pigment myoglobin with hemoglobin. Myoglobin has a half-saturation pressure of about 6 mm Hg, and that of human hemoglobin is about 24 mm Hg. Myoglobin remains complexed with oxygen until the oxygen levels of the muscle are lowered considerably - a time at which the muscle requires oxygen. The oxygen dissociation curve of myoglobin is hyperbolic, whereas that for hemoglobin is sigmoid. Myoglobin contains only one chain and one heme group and can combine with only one oxygen molecule. The sigmoid oxygen dissociation curve of hemoglobin results from the combination of hemoglobin with four oxygen molecules. As is desirable in an oxygen storage pigment, myoglobin hangs onto all of its oxygen at oxygen tensions where hemoglobin is 98 percent dissociated. The hemoglobin equilibrium with oxygen has been described mathematically as:

$$\frac{y}{100} = \frac{KP^n}{1 + KP^n}$$

where y is the percent saturation with O_2; P is the partial pressure of oxygen; K is the equilibrium constant of the reaction; and n is a measure of the interaction between heme groups. After rearranging and taking the logarithms of both sides:

$$\log \frac{y}{100 - y} = \log K + n \log P$$

A plot of log $[\frac{y}{100 - y}]$ against log P yields a straight line whose slope is n. The intercept on the log $[\frac{y}{100 - y}]$ axis gives the value of K. For myoglobin, which contains only 1 heme group, n = 1. For hemoglobin, between 10 and 98 percent saturation, n = 2.7.

In hemoglobin the interaction of one heme group with an oxygen molecule increases the affinity of the other groups for oxygen. Since the three-dimensional structure of hemoglobin shows that the heme groups are relatively separated, it is thought that combination of oxygen with one heme group affects the total protein structure, thus making other heme groups more available for oxygenation.

Oxygen dissociation curves are also affected by the pH and by the blood P_{CO_2}. Low pH or high P_{CO_2} both shift the curve to the right (in most cases). This is the Bohr effect and is important in the functioning of oxygen transport pigment. In active tissues the P_{CO_2} is elevated, and the blood is slightly more acid. Under these conditions the shift of the oxygen dissociation curve to the right under the influence of pH facilitates oxygen unloading and therefore increases the amount of oxygen available to active cells. The mechanisms by which P_{CO_2} and pH affect the dissociation curve appear to be different, although high P_{CO_2} decreases blood pH. Increases in temperature also shift the oxygen dissociation curve to the right.

Carbon Dioxide Transport

Carbon dioxide produced by cells is transported to the lungs or other respiratory surfaces for excretion. Only small amounts of CO_2 are in solution in the blood because of reactions that occur in the plasma and in the erythrocytes of vertebrates.

CO_2 dissolved in water forms carbonic acid, H_2CO_3, a reaction facilitated by the enzyme carbonic anhydrase found in the blood of many animal species. In vertebrates carbonic anhydrase occurs primarily in erythrocytes. Carbonic acid can dissociate into H^+ and HCO_3^- and these ions then can enter into the acid-base regulating mechanisms to be discussed in the next section.

Some carbon dioxide (15 to 20 percent) reacts with hemoglobin:

$$HbNH_2 + CO_2 \rightleftharpoons HbNHCOOH \rightleftharpoons HbNHCOO^-$$

where NH_2 represents amino groups of the amino acids of the hemoglobin. The complex is known as carbaminohemoglobin.

At the respiratory surface CO_2 diffuses from the blood to the atmosphere or aqueous medium. Carbonic anhydrase catalyzes the reverse reaction from that given above and forms CO_2 that can be eliminated. The direction of the reaction depends on the P_{CO2}. Where the CO_2 pressure is low, as in the region of the respiratory exchange surface, H_2CO_3 is split into H_2O and CO_2, and the latter moves down its concentration gradient to the exterior.

It may be noted that the interactions of P_{CO2} and P_{O2} levels at different sites in the animal are coordinated and intermeshed in such a way that oxygen diffusion to tissue cells is facilitated just where oxygen is most needed, while oxygen intake from the environment is also facilitated. Similarly, carbon dioxide removal from the tissues to the blood and then from the blood to the external medium is facilitated by the actions involved in oxygen transport. The two systems are interwoven to fill the requirements of the organism.

Acid-Base Balance

When CO_2 diffuses into the blood and forms carbonic acid, the dissociation of the latter forms H^+ and HCO_3^- . The hydrogen ions, unless buffered, would cause dangerously low pH levels in the animal. They are taken up by proteins of the blood in exchange for metal ions such as K^+ and Na^+. $NaHCO_3$, $KHCO_3$, and phosphates are formed in the blood. Thus plasma components act to buffer the blood pH and prevent it from decreasing. These reactions also reduce the further formation of HCO_3 from carbonic acid and keep CO_2 flowing into the system from the tissues. It should be noted that this is based on the maintenance of adequate concentration gradients. When a given substance is converted to a new compound, the concentration of the latter does not influence the diffusion of the original substance.

Erythrocyte hemoglobin also aids in buffering the blood pH and allows more CO_2 to enter the blood from body tissues. Then CO_2 diffuses into the erythrocytes from the plasma and is converted to H_2CO_3 by erythrocyte carbonic anhydrase. Again, the carbonic acid dissociates to form H^+ and HCO_3^- within the cell.

At the tissue level, hemoglobin gives up its oxygen and may then combine with H^+: $Hb^- + H^+ = HbH$. This reaction is of extreme importance in maintaining the pH of the cell at a constant level. The structure of hemoglobin is such that when it gives up oxygen it is converted at the right time and the right place to a substance capable of combining with the hydrogen ions formed as a result of CO_2 diffusion from the tissues.

The CHO_3^- ion concentration within the erythrocyte increases because of the reactions just described, and bicarbonate diffuses out to the plasma. The erythrocyte membrane is not permeable to cations, so in order to preserve electrical neutrality, for every HCO_3^- ion that diffuses into the plasma, a Cl^- moves into the erythrocyte, where it associates with an alkali metal ion, usually Na^+ or K^+. This is known as the chloride or Hamburger shift.

At the respiratory surface all of these reactions reverse because CO_2 can diffuse into the external medium. When hemoglobin combines with oxygen, it releases hydrogen ions (the Haldane effect). The oxygenated hemoglobin associated with the alkali metal ions that were originally associated with the $C1^-$. Chloride ions diffuse out of the cell, and HCO_3^- diffuses back into the cell to react with the H^+. Carbonic acid is formed and then rapidly converted by carbonic anhydrase to CO_2 and water. The CO_2 follows its concentration gradient and diffuses out of the cell, through the plasma, into the external environment.

It is to be noted that the operation of this system depends upon the presence of proper levels of protein, alkali metal ions, Cl^-, HCO_3^-, and phosphate. One function of the kidney is to maintain appropriate levels of Na^+ and K^+ in the body fluids. An important need for relatively high levels of these ions arises from their involvement in the buffering system that maintains blood pH at its proper level even while the acid waste product CO_2 is being transported. The kidney tubules have enzyme systems and buffering mechanisms that trade waste products such as NH_4 and H^+ for ions such as Na^+ and K^+. There is also an exchange of HCO_3^- or $C1^-$. pH is maintained while, simultaneously, needed ions are retained.

Proteins have many ionizable side groups because of the nature of their constituent amino acids, and a molecule such as hemoglobin is capable of reacting with 20 to 50 positively charged ions. That is, the buffering capacity of proteins is relatively high. When blood proteins are low in concentrations because of disease or metabolic upsets, H^+ begins to accumulate, lowering the

Table 3-4. Subgross Pulmonary Anatomy Types: Man and Common Laboratory Animals

	TYPES			
	I	II		III
Species	Pig (Rat) (Mouse)	Monkey, Cat, Dog Guinea Pig, Rabbit Rat, Mouse, Ferret		Human (Rabbit)
Lobulation	Extremely Well Developed	Absent		Imperfectly Developed
Plura	Thick	Thin		Thick
Arterial Supply to Pleura	Bronchial Artery	Pulmonary Artery		Bronchial Artery
General bronchovascular relationships (in comparative diameters)	PV++ BR++ BA++ PA++	PV+++ BR++ BA++ PA++	Distal PV++ BR++ BA 2 (+) PA++	Proximal +++ ++ 2 (+} ++
Intrapulmonary Termination of Bronchial Artery	Distal Airway	Distal Airway		Distal Airway and Alveoes
Terminal Bronchioles	Present Predominant Distal Airway	Absent		Present
Respiratory Bronchioles	Infrequently Observed Extremely Poor Development	Present Very Well Developed		Present Poorly Developed
Bronchial Artery Pulmonary Artery Shunts	Present (Not Demonstrated in the Pig)	Not Demonstrated		Present

( ) = species fits this category to a lesser degree.

blood pH and resulting acidosis. Low blood pH will, in turn, create a low pH in body fluids and cells (alkalosis). Unless such a condition is rectified, death will ensue because cells can withstand only relatively small changes in pH. In mammals, for example, changes in blood pH greater than a few tenths of a pH unit result in death.

Table 3-4 presents a comparison of the structural characteristics of the lungs of man and the common laboratory species. Based on the majority of such characteristics, these species are grouped into three major families. Note that only the rabbit is in the same class as humans - but differs in other ways which limit its usefulness as a model for humans. Note that there really is no ideal or best animal model for studying inhalation in humans.

Complications of Temperature

One of the reasons that combustion toxicology is more complicated than the broader field of inhalation toxicology is that biological systems (humans and animals) are subject to an additional variable, stress (*i.e.*, heat). Increased environmental and internal temperatures represent a source of additional stress that has been largely excluded from consideration in current test system design.

Influence on Toxicity

Weike (1973, 1976) has reviewed the effects of temperature on a range of animals and their response to chemical agents. He pointed out the important differences between laboratory animals (mouse, rat, guinea pig, rabbit) and man in their mode of body temperature regulation when exposed to cold and heat due to experimentally enforced restrictions of behavioral temperature regulation in the animals. Laboratory animals in confinement adapt to heat and cold metabolically while man responds by controlling heat loss by insulative adaptation. Although humidity is usually treated separately, it contributes significantly to the thermal environment as latent heat. In his review, he concluded that the sensitivity scale for toxicity (from high to low) in mice exposed to amphetamine and housed singly or in groups and at various ambient temperatures, was as follows: warm grouped mice; warm single mice; cold grouped mice; and cold single mice. Three to nine mice grouped in a cage at 26°C (room temperature) showed a fourfold increase in amphetamine toxicity, which was eliminated when the mice were cooled by a rapidly moving airstream. This should obviously be taken into consideration in performing combustion studies. These differences in toxicity appeared to be due to differences in heat production and heat loss for single and grouped mice. Grouped mice will always aggregate to minimize heat loss in the cold and separate to maximize heat loss in the heat.

Fuhrman and Fuhrman (1961) analyzed the patterns of toxicity for several agents in mice and rats exposed at different temperatures. The first of these patterns was the V- or U-shaped response curve, with minimum toxicity around thermal neutrality and increasing toxicity at both lower and higher temperatures. The curve for metabolic rate as a function of temperature is also V- or U-shaped so that the toxicity pattern parallels metabolic rate. It appears that this type of toxicity pattern is determined by metabolic rate rather than by resetting body temperature. The groups of agent exhibiting this pattern also include those that produce hypothermia, such as chlorpromazine and reserpine.

This pattern is characteristic of many centrally acting agents affecting the thermoregulatory center, including hypnotics and neuroleptics such as phenothiazines, monoamine oxidase inhibitors, rauwolfia alkaloids, cholinergic blocking agents, morphine, salicylates, and other antipyretic drugs, as well as the digitalis glycosides. This same pattern was also reported for parathion administered intraperitoneally to mice (Baetjer and Smith, 1956).

Fuhrman and Fuhrman (1961) reviewed these results, as well as those for the anticholinesterase sarin, administered intramuscularly to Rhesus monkeys. Both of the anticholinesterase agents, parathion and sarin, were reported to be more toxic at higher environmental temperatures. Experiments in which monkeys were exposed to sarin vapor in a box that excluded their heads also demonstrated a greater toxicity at 38°C than at 25°C. Since muscle and skin blood flow are also greater at 38°C, it was suggested that an increased rate of absorption was responsible for the increased toxicity at higher environmental temperatures.

Funckes *et al.* (1963) showed that the rate of absorption of parathion in man was temperature related and that the excretion of p-nitrophenol varied directly with temperatures from 14.4 - 40.5°C. In field use, the temperature effect is probably more apparent by percutaneous than by inhalation exposure (Durham *et al.*, 1972). However, the inhalation toxicity of tinox, dimethoate and methyl parathion are profoundly affected by environmental temperature (Grigorowa and Binnewies (1973).

A second pattern of toxicity, a linear response curve pattern, can be seen with increasing toxicity at increasing temperatures. This is characteristic of sympathomimetic drugs such as amphetamine and methamphetamine, epinephrine and norepinephrine, procaine, cortisone, analeptics such as picrotoxin and antihistamines such as diphenhydramine.

Finally, an angled response curve pattern can be seen with constant values of toxicity at low temperatures up to the thermoneutral zone and increasing toxicity above the upper critical temperature. The occurrence of this type of response is debatable, and agents in this group may prove to belong to the V- or U-shaped or the linear response curve pattern. Although the toxicity of many chemicals is influenced by temperature, these effects are not fully understood. They are, however, associated with metabolic adaptation and physical and behavioral temperature regulation. In animal experimentation where behavioral temperature regulation can be largely eliminated, the effects are impressive, whereas in man, where behavioral temperature regulation is operative, these effects are mild.

Keplinger *et al.* (1959) studied the effect of environmental temperature on the approximate lethal dose of 58 agents administered intraperitoneally to rats. They demonstrated that both central nervous system stimulants and depressants are much more toxic at higher temperatures (36°C) than at room temperature (26°C). Nearly all of the compounds were more toxic at 36°C than at 26°C. Chlorpromazine and promazine were most toxic at 8°C, and more toxic at 36°C than at 26°C. Strychnine, however, was equally toxic at 8°C and 36°C, but much less toxic at room temperature. Although signs of intoxication were not markedly influenced by environmental temperatures, the onset of action was more rapid and the period of survival was shorter in rats exposed at 36°C. Central nervous system stimulants, caffeine and pentylenetetrazol were equally toxic at cold and room temperatures, while central nervous system depressants, such as paraldehyde and pentobarbital, and solvents such as benzene, kerosene, toluene, and the nitro-olefins were more toxic in the cold than at room temperature.

Phenol derivatives, including pentachlorophenol, 2,4-dinitrophenol and 4,6-dinitro-o-cresol, which causes a marked increase in metabolic activity (with resultant fever), were less toxic in a cold environment. This has been reported for pentachlorophenol and for dinitrophenol, and is the reason why cold baths or ice packs are recommended for the treatment of acute intoxication by these compounds. It is well known that the phenothiazine derivatives promazine and chlorpromazine cause hypothermia, and this may explain why they were most toxic at 8°C.

By what mechanisms environmental temperature affects the toxic doses of compounds is not obvious and may differ with each compound. Since death was the endpoint of the test, it is difficult to determine accurately whether the stress of a changed environmental temperature rendered a rat more susceptible to the compound or the compound rendered a rat more susceptible to heat or cold. Environmental temperature can alter normal body functions so that the rate of respiration, absorption, diffusion, distribution, or metabolism of a compound would be changed. This can also influence a compound's rate of reaction with, or affinity for, receptor sites in the body.

The mechanisms of metabolism or detoxification of various compounds in the rat are not fully known and may differ from those of man. The mechanism by which body temperature is controlled also differs in the rat and man. These factors must be taken into consideration in extrapolating rat data to man. The rodent lacks sweat glands and, therefore, does not lose water by the same mechanism as man. However, when a rat is exposed to a stagnant, hot, and humid atmosphere, it's fur becomes quite wet. This suggests that the rat, in addition to imperceptible water loss, is capable of covering its surface with a film of moisture so that the physiological function of the skin regarding water elimination may be comparable to man. It may therefore, be assumed that the effects of therapeutic doses of drugs, as well as maximum allowable concentrations of chemicals, are influenced in man by environmental temperature.

Physiological or Stress Response

Whenever animals are exposed to heat stress, a number of systems respond in an effort to rid the body of the thermal load. Peripheral vasodilation occurs; as peripheral resistance falls, heart rate and cardiac output increase. If temperature gradients between the skin and the surrounding air are adequate, convective heat dissipation increases to restore the balance. If not, or if the thermal load is large, sweating occurs. Evaporation is an extremely effective method of heat dissipation in man (but not rodents) because of the high latent heat of vaporization of water. For evaporation to occur, however, the air surrounding the body must have a vapor pressure below the saturation vapor pressure at skin temperature.

If these mechanisms for heat loss are not adequate to keep pace with the rate of heat input to the body, heat storage must occur; the core temperature will rise. A number of symptom complexes are seen when man encounters heat stress beyond his capacities.

Heat stroke is the most serious of these syndromes. Its presence is made known by anxiety, irritability, visual disturbances, delirium, collapse, coma, and death if treatment is inadequate. In man, the skin is dry, hot, and flushed. The rectal temperature in humans is very high (40 to 43°C). Cessation of sweating in the presence of continued thermal stress, probably owing to failure of the central thermoregulatory centers, allows a very rapid increase in the rate of heat storage.

Heat stroke constitutes a true medical emergency. Removal of the individual from the source of heat stress, followed by vigorous attempts to cool the body, are the first measures to be instituted. Rapid cooling has to be used, even though skin cooling induces peripheral vasoconstriction, which inhibits conduction of core heat to the surface.

As the rectal and core temperatures begin to fall, it becomes necessary to moderate treatment so that one does not induce hypothermia, since the thermoregulatory failure which allows heat to be retained will also fail to protect against excessive heat loss. In essence, someone or something must take over the task of maintaining thermal homeostasis until the individual recovers that capacity. Individuals who have had heat stroke may tolerate intense heat poorly thereafter, because of permanent damage to the thermoregulatory centers.

Heat exhaustion is characterized by weakness, hypotension and elevated temperature. The skin is not dry, however; often the patient has been sweating vigorously. The condition is probably caused by moderate dehydration and salt imbalance. Removal from heat and treatment with fluid and electrolytes (the intravenous route may be necessary if the patient is nauseated) usually produce prompt recovery. There are no long-term sequelae.

Mechanisms of Toxicity

As with exposures by any of the other routes, inhaled chemicals (no matter what their physical state) can cause toxic effects in one of two broad categories - local effects or systemic effects. The broad mechanisms involved and the ways in which effects resulting from acute exposures are expressed need to be considered.

First, however, it should be remembered that certain functional peculiarities make the respiratory tract uniquely vulnerable to toxic injury. Although situated deep within the body, the lung is in immediate contact with the external environment at each breath and this contact is renewed at least 20 times per minute in humans (more often in rodents). During strenuous exercise or labor, the frequency and depth of respiratory effort, and hence air intake, is markedly increased, leading to potentially greater toxic exposure. Secondly, the alveolar lining cells have a large surface area and thin cytoplasm, with few organelles. This combination of large absorbing surface and modest defensive ability renders the thin surface epithelial (type I) cells vulnerable to toxic injury. Thirdly, the alveolar surface is extremely large (thirty times that of the skin), and its abundant thin-walled capillaries lie in intimate contact with the epithelium. The alveolar surface represents the largest area in the body at which the blood comes into virtually direct contact with the external environment. Consequently, toxic fumes are capable not only of injuring extensive areas of alveolar tissue and adjacent structures but also of initiating rapid absorption and distribution throughout the body to injure specific target organs, or the body as a whole.

Any foreign substance can product irritation at the point of contact with the respiratory system. Gases and vapors tend to be more chemically reactive and therefore more irritating, but particulates can serve as “carriers” for other phases (liquid or gas) absorbed to their surface.

In the lung, these irritants can interact with pulmonary membranes to produce edema, pleural effusion, and a hyperemic reaction. With the breakdown of the pulmonary barrier, the flow of fluid and protein becomes altered, leading ultimately to the loss of the ability to appropriately oxygenate the blood. The inflammatory process is essentially the same within the lung as elsewhere in the body, but selected portions of that process may be affected, due to differences in the solubility, boiling point, and volatility of the irritant chemical. The solubility of a compound is probably the most important factor in determining its site of action. Highly soluble materials such as ammonia and hydrogen chloride primarily affect the upper respiratory tract, producing rhinitis and related inflammatory changes. Compounds of intermediate solubility, such as ozone, affect both the upper respiratory tract and the pulmonary tissue, while less soluble materials such as nitrogen dioxide and phosgene affect primarily the deep lung.

Irritants can produce changes both at the site of contact and systemically. Those which act only locally at the point of contact are considered primary irritants. These exert little systemic toxicity since they are either metabolized to products which are nontoxic, or the local effects far exceed any systemic reaction. Examples of primary irritants are hydrogen chloride and sulfuric acid.

Another type of primary irritant is characterized by reactive materials such as the nitrogen mustards. These compounds are extremely toxic if absorbed but produce such a rapid and severe degree of irritation that death occurs before systemic poisoning can be produced. Secondary irritants are agents which have a significant irritating effect on mucous membranes but, following absorption, have a more profound systemic effect. Hydrogen sulfide, acetic acid and many organic compounds fall into this category. There are materials which are irritants but also produce systemic injury; hydrogen fluoride, for example, produces severe injury at the point of contact but also produces kidney damage.

Inhalation of a wide variety of particulates produces pulmonary disease. Pneumoconiosis is a term used to designate a fibrotic condition of the lung caused by inhaled dust. The character of the fibrosis depends on the type of dust inhaled, with the more important agents being crystalline silica (silicosis), asbestos (asbestosis), talc, and coal that contains silica (see Kennedy, 1988). Deposition of metals and their compounds in the lung produces a variety of disease states. The pulmonary response to particulates is determined by factors such as: (1) the nature of the particles, (2) their site of deposition in the respiratory tract, (3) the amount of deposition in the lung, (4) the exposure period, and (5) individual variation and immunologic status. Lee (1985) described five categories based on the lung tissue reaction to inorganic (mineral) dusts. These include the macrophage reaction produced by so called nuisance dusts such as soot, iron, and titanium dioxide, foreign body

Table 3-5. Responses Of The Lungs To Toxic Injury

1. **Alveoli**	**Type**	**Responses**	**Causes**
Acute	epithelial injury	a) simple repair	O_2
		b) exudation and fibrosis	NO_2, SO_2, Cd^{++}
	endothelial injury	thrombosis	O_2, O_3
	interstitial injury	fibrosis	proteolytic enzymes
Chronic	granulomatous injury	tuberculosis beryllium *etc.*	
	immunologic injury	farmer's lung	
2. **Bronchi**			
Acute	mucosal ulceration metaplasia		
Chronic	atypia dysplasia neoplasia (different patterns of cancer)		
3. **Systemic damage**	central nervous system target organ toxicity		

granuloma produced by beryllium and talc, sarcoid-type granuloma produced by beryllium, collagenized fibrosis of both the diffuse (from hard metals, asbestos) and nodular (from quartz and silica) forms, and neoplasia produced by nickel and asbestos.

The range of local effects limited to the lungs is illustrated in Table 3-5.

A broader (but less detailed) characterization of toxic effects resulting from respiratory tract exposure includes the following:

I. Systemic Effects
  A. Lungs (see Menzel and McClellan, 1980 for a discussion of five categories of actions on lungs).
  B. Any other possible target organ (see Lu, 1985).
II. Local Effects
  A. Local irritation.
  B. Cellular damage and edema.
  C. Fibrosis and emphysema.
  D. Immunologic responses.
  E. Upper respiratory tract effects.
    1. Hyperemia
    2. Deciliation (Cobb, 1981)

3. Goblet cell hyperplasia

The special case of deaths due to fire, meanwhile, has six major causes.

1. Oxygen deficiency (anoxia).
2. Poisoning from carbon monoxide.
3. Heat destruction of tissues and thermal shock.
4. Irritation of the respiratory tract by various gases and aerosols.
5. Development of fear.
6. Reduced visibility due to smoke preventing escape.

Absorption and Distribution

Inhalation occurs when the gaseous or airborne phase of a material is drawn into the respiratory tract. Characteristics of the inhaled material and of the person or animal inhaling it will determine how much material is actually taken in, how deeply it penetrates the tract, and how well it is absorbed.

The respiratory tract provides an enormous ($7.5m^2$ in the rat - 10 times this in humans) and efficient absorption surface for potentially toxic materials in many forms.

Gases penetrate the respiratory tract epithelium with great rapidity. These substances have small molecular sizes and low lipid-to water partition coefficients, but often their high lipid solubility greatly enhances absorption.

Very little quantitative information is available concerning the pulmonary absorption of organic solids. Although many drugs and other chemicals appear to be absorbed readily when inhaled as sprays, aerosols or dusts, the observations are based on the appearance of pharmacologic action or the appearance of materials in blood or urine; consequently, rates of absorption and relative rates of absorption are not as well known.

An indication of a lipid-like character for the pulmonary membrane has been supplied by a quantitative study with the isolated, perfused dog lung (Taylor *et al.*, 1965). it was shown that the lipid-soluble compound dinitrophenol crosses the pulmonary epithelium more than one hundred times faster than a lipid-insoluble substance such as glucose.

Quantitative studies with the rat lung *in vivo*, such as those by Ennha and Schanker (1969), suggests a lipid-pore nature for the pulmonary membrane. In this work, 0.1 ml of drug solution was administered to anesthetized rats through tracheal cannula. After various periods of time, the lungs and trachea were removed and assayed for remaining drug. Although lipid-insoluble substances such as p-aminohippurate and sucrose were absorbed fairly readily, much more rapid rates of absorption were seen with lipid-soluble compounds such as aniline and procaine amide. The presence of membrane pores suggested by the observation that inulin, a large, lipid-insoluble molecule (MW 5,000) was absorbed at a significant rate, and the smaller sucrose molecule was absorbed even more readily. The reader is referred to Munson and Eger (1971) for a more detailed discussion of the absorption of gas phase materials, which primarily diffuse across the alveolar membrane and enter the circulation.

Chemical moieties may also be inhaled as aerosols, and many toxic substances enter the body in this way. Aerosols are liquid or solid particles so small that they remain suspended in air for a long time instead of sedimenting rapidly under the force of gravity. Aerosols have probably been the most overlooked hazardous portion of the fire atmosphere.

Stokes' law is the primary theoretical statement of the behavior of a particle in a gas and is stated as: f = 6 nrv where f is force in dynes; n is the viscosity of air at 20°C and atmospheric pressure and is equal to 1.9×10^{-1} g sec^{-1} cm^{-1}; r is the radius of sphere in centimeters; and v is the constant velocity of movement when force f is balanced by the viscous drag.

For a sphere moving under the force of gravity, f = mass × g, where g = 980 cm sec^{-2}; and assuming unit density.

$$f = \frac{4}{3} r^3 g$$

When the viscous drag opposes the force of gravity so that the rate of fall is constant,

$$6\ nr_v = \frac{4}{3} r^3 g$$

$$v = \frac{D^2 g}{18n}$$

where D is the diameter of the sphere in centimeters and $v = 2.87 \times 10^2 5D^2$.

Particles below about 10 microns in diameter are of interest in humans,because they achieve deep lung penetration, though it is frequently overlooked that in the rat (the primary animal model) this cut-off is 3 microns.

Once deposited, a particle may be cleared from the respiratory tract (primarily from the pulmonary region) as the physiologic mechanism for clearance from the NP and TB regions - coughing and swallowing - are more straightforward and effective by various means which include ciliary action (the mucocilliary "elevator"), alveolar macrophages, phagocytosis (into the lymphatic system) and solubility. These clearance mechanisms are, with the exception of solubility, adversely affected by particle cytotoxicity and fiber length (above 50 microns, there is no clearance). A mucous blanket, propelled toward the head by ciliary movements, covers the upper respiratory tract down to the terminal bronchioles, and impacted aerosol particles are cleared by this mechanism.

Particles that deposit in the alveolar sacs must first be transported up to the mucous layer to be cleared. Particles in the alveoli are also ingested by phagocytic cells. The efficiency of clearance of solid particles from the lungs is remarkable (not more than a minute fraction of all the mineral dust inhaled during a lifetime is retained in the lungs). However, a small decrease in the clearance capacity of the lungs could cause a marked increase in the amount of retained particle matter; this may be a factor leading to the development of pneumonoconiosis in miners exposed to silica dusts.

In addition to systemic toxicity, particles deposited in the pulmonary region can have other local biological effects. These include synergism with gases and vapors, irritation, fibrosis and infection.

Particle size strongly influences the rate at which material is absorbed through the alveolar epithelium, probably because of the greatly increased relative surface area for solubilization as particle diameter becomes smaller. For example, particles of uranium dioxide larger than 3 microns had no toxic effect whatsoever when introduced into the trachea in rats, but the same relatively insoluble material was absorbed into the circulation and caused kidney damage when smaller particles were employed.

Likewise, numerous metals (such as lead) in aerosol form are extremely well absorbed. Active chemical moieties in aerosol form can elicit extremely rapid responses when inhaled, due to the minimal separation between the gas phase and the systemic circulation in the deep lung. Histamine given to a guinea pig or dog as an aerosol can cause bronchiolar constriction and asphixia within a minute.

Particles larger than 2 microns in diameter probably do not reach the alveolar sacs of humans. With aerosols of small particle size, the amount of chemical reaching the alveoli may be large; and since the rate of absorption into the bloodstream is much more rapid at the alveolar sacs than elsewhere in the pulmonary tree, the systemic absorption of the drug may be appreciable.

Vapors have some special characteristics and behaviors which set them apart from simple gases. Because they are the gas phase of materials that are predominantly liquid at ambient temperatures (25°C) and pressures (1 atmosphere) and because combustion generates sufficient concentrations of the gas phase (to meet our experimental needs) by subjecting the test material to increased temperatures, careful consideration must be given to keeping the material as a gas phase. Subjecting an air stream saturated with the vapor of a chemical to cooling will, at the least, cause a reduction in vapor phase concentration (as some of the chemical condenses out as a liquid along the walls of the air handling or exposure system) and may lead to the inadvertent generation of an aerosol as the vapor condenses out as a mist. Of course, this can and will also happen in a real fire situation.

There is also a special problem for vapors which are mixtures of chemical moieties. Achieving and maintaining the desired representative concentration of all the components in a vapor mixture (as opposed to a unitary vapor) can be a complex problem.

The major toxic gas formed in most fires is carbon monoxide. The major pathophysiologic disturbance in cases of carbon monoxide poisoning is that of tissue hypoxia. Tissue hypoxia occurs as a consequence of the extraordinary avidity of carbon monoxide for hemoglobin (more than 200 times that of oxygen). Carbon monoxide not only displaces oxygen from hemoglobin but also interferes with the release of oxygen by the pigment to the tissues.

Carbon monoxide poisoning is often rapidly fatal. Without an emission-control device, lethal concentrations of the gas can accumulate inside a car or a small garage in 15 to 30 minutes (McBay, 1965). The clinical manifestations of carbon monoxide poisoning depend primarily on the duration and intensity of exposure (Burney *et al*, 1982). Sudden, high concentrations of carboxyhemoglobin in the blood result in rapid loss of consciousness, coma, seizures, and death. When poisoning is slow, a sequence of influenza-like symptoms usually ensues: a throbbing headache and irritability develop, followed by dizziness, weakness, impaired vision, ataxia, nausea, vomiting, syncope, and eventually, unconsciousness. Such variables as physical activity, preexisting anemia, vascular disease, and the state of respiratory function influence the severity of poisoning. Cherry-red discoloration of the skin appears in only a minority of victims.

Mild disturbances are associated with blood concentrations of carboxyhemoglobin up to about 20 percent. Saturation of from 30 to 50 percent of the hemoglobin molecules with carbon monoxide produces severe syndromes, and blood concentrations of carboxyhemoglobin higher than 50 percent are often fatal. Metabolic acidosis that occurs despite the presence of hyperventilation ($P_{CO2} < 49$) is an ominous sign (Larkin *et al.*, 1976).

Heavy smokers may have carboxyhemoglobin levels in the range of 7 to 10 percent.

Irritant gases and vapors produce inflammation in tissues as a result of direct contact, usually with the surface of the skin and mucous membrane. In fire situations, gases and vapors affecting the eyes and respiratory tract may cause tissue injury in themselves, or may make it difficult to see and/or breathe, with the result that exist may be delayed until victims are overcome by other injuries, such as CO poisoning or burns. Irritants may also cause post-exposure effects that further strain an already injured respiratory tract.

Sensory irritants are irritants of the upper respiratory tract. These chemicals inhibit respiration and cause one or more typical reflex actions, associated with stimulation of trigeminal nerve endings, such as a burning sensation in the nose and/or eyes, moist facial skin, sneezing, coughing, and tears. Typical sensory irritants are hydrogen chloride (HCl), sulfur dioxide (SO_2), acrolein, and ammonia. Pulmonary irritants of the lower respiratory tract typically increase respiratory rates as a result of stimulation or sensitization of various nerve endings of the lower respiratory tract, and such action is usually without immediate effect in the form of pain. True pulmonary irritants are ozone, nitrogen dioxide, and phosgene. Most pulmonary irritants also exert an effect as sensory irritants. Nitrogen oxides (NO_X) are the most likely pulmonary irritants in fire situations. Nitrogen dioxide may cause pulmonary edema, which is serious, in itself or as a contributing factor with other fire injuries.

Other types of irritant gases and vapors are bronchioconstrictors and respiratory irritants. The former are chemicals that induce increased resistance to airflow in the lungs. Typical examples are sulfur dioxide (SO_2), ammonia (NH_3)and sensory allergens such as foreign protein, toluenediisocyanate (TDI) or histamine. The term "general respiratory irritants" is used to cover chemicals that, when inhaled, can act as sensory irritants, pulmonary irritants, and bronchioconstrictors. Typical examples are chlorine (Cl_2) and ketone.

In addition to oxides of nitrogen, probably the most serious respiratory irritants in ordinary fire situations are the aldehyde gases found in dense smoke. Acrolein (CH_2CHCHO) is a good example.

In mixed fuels or natural polymer fires, smoke particulate is a thick, oily aggregate of carbon particles and adsorbed combustion products that are usually irritants. Massive exposure may cause direct mechanical plugging of the respiratory passages, but more commonly, smoke particulate is irritating to at least a minimal degree. The nature of the smoke particles generated may vary with the temperature of the fire. Smoke particles may be small enough to be inhaled deeply into the lungs. These particles can convey adsorbed irritants. Smoke can also reduce vision and lead to panic. Smoke obscuration is a definite hazard in itself, since exits cannot be readily located. In some cases, the addition of fire retardants results in a decreased tendency to ignite, but an increased tendency to smoke when ignition finally takes place. This "trade-off" in properties can be responsible for loss of lives.

Miscellaneous gases and vapors may be adsorbed on carbon smoke particles, or may be airborne in the first effluent. Carbon dioxide from the complete oxidation of carbon atoms may be the most common product in fires, particularly in hot, well-ventilated ones. CO_2 is toxic by itself, but only in high concentrations. However, CO_2 stimulates respiration. A 3% concentration doubles lung ventilation. Depending on the chemical structure of the fuel and the particular fire situation, other gases with anesthetic or drug-like properties, such as propane or benzene, may be released in relatively small amounts in combustion products.

Pulmonary Irritation

Chemicals irritating the sensory apparatus of the upper respiratory tract can be identified by measuring respiratory rate to detect a reflex-induced decrease (Alarie, 1966; Alarie *et al.*, 1974, Armdur, 1957). The basic design uses an apparatus which detects changes in thoracic body cavity volume that can determine the depth and rate of respiration. The animal, usually a rat or mouse, is placed into a whole-body tube (plethysomograph) connected to an inhalation chamber such that only the nose protrudes into the exposure area. Animals are then exposed to various concentrations of the agent and a plot of respiration rate against concentration is made. Dose-response curves and minimum effect levels can be determined, the data usually being expressed in terms of the RD_{50} (concentration required to lower the respiration rate by 50%). The method, as described below, is quite simple and detects the effects of irritation at concentrations where no associated pathological modifications occur.

The use of this method has also been demonstrated for a variety of airborne chemical irritants in numerous species: cats, dogs, mice, rats, rabbits, guinea pigs, and humans. This reflex reaction has been exploited as a quantitative measure for evaluation of sensory irritation to airborne chemicals. It consists of measuring the percentage decrease in respiratory rate during exposure to various concentrations of a given gas, vapor, or aerosol. The animals serve as their own controls since the respiratory rate before exposure offers a base line. A dose-response curve is obtained by plotting the percentage decrease in respiratory rate relative to control values during the exposure interval against the logarithm of the exposure concentration. Alarie has used this as a portion of his combustion toxicity testing system (Alarie *et al.*, 1974).

LEXICON

Acute Toxicity

The effects following a single exposure or dose of a toxicant. The effects may be observed immediately or after a delay of hours, even days.

Anoxia

A complete lack of oxygen in the tissues, *e.g.*, the blood or the brain. The term hypoxia is preferred, since there is rarely a total absence of oxygen.

Biological Assay

A term for the use of a biological system to detect and/or measure a biologically active material. In the fire context, it refers to the use of animal exposures, rather than chemical analysis, to determine the toxicity of a fire effluent atmosphere.

Brochoconstriction

Constriction of the airways in the lung due to the contraction of the smooth muscle in the airway walls. It is a response either to the direct action of an agonist or its indirect stimulation of irritant receptors.

Chronic Toxicity

Toxicity resulting from multiple doses of, or exposures to, a toxicant over an extended period of time.

Concentration

The amount of a contaminant in the atmosphere per unit of volume, usually quoted as weight/volume (mg/L or g/m^3) or volume/volume (ppm or %).

Delayed Toxicity

A toxic effect which is not manifested until a period of several hours, days or, in some cases, weeks after exposure.

Dose

The body burden of a toxicant. In inhalation toxicity the dose can only be estimated, *e.g.*, by multiplying the atmosphere concentration by the time duration of exposure.

EC_{50}

Effective concentration 50%. A concentration statistically calculated to cause an effect in 50% of the animals exposed for a defined period of time. In the fire effluent context it usually refers to incapacitation.

ECT_{50}

The mathematical product of time of exposure and concentration of a toxicant statistically calculated to cause an effect in 50% of the animals.

Exposure Dose

An approximation of dose which takes into account the amount of material consumed per unit of time, the volume into which it is diluted and the duration of the exposure. It is expressed in $mg.L^{-1}.min$ and is determined using weight loss values obtained during the course of combustion.

Hyperventilation

Abnormally increased rate and depth of breathing, usually in response to tissue hypoxia, increased carbon dioxide concentration or stimulation of the irritant receptors.

Hypoxia

A reduced oxygen concentration in the tissue. This can be caused in four ways: (1) Anaemic Hypoxia, in which the arterial pO_2 is normal but the amount of haemoglobin available to carry O_2 is reduced (*e.g.*, by formation of carboxyhaemoglobin following exposure to CO); (2) Histotoxic Hypoxia, in which the amount of O_2 delivered to the tissue is adequate but, due to the action of a toxic agent (*e.g.*, HCN), the tissue cells cannot make use of the O_2 supplied; (3) Hypoxic Hypoxia, in which the pO_2 of the arterial blood is reduced as a result of low atmospheric oxygen concentration or factors such as impairment of gas exchange in the lung due to bronchoconstriction, respiratory tract damage or disease; and, (4) Ischemic Hypoxia, in which blood flow to a tissue is so low that adequate O_2 is not delivered despite normal pO_2 and haemoglobin concentration.

Incapacitation

Inability to perform a task, presumably related, in a fire context, to escape.

Irritation (Pulmonary)

The action of irritants on the lower respiratory tract which may result in breathing discomfort (dyspnoea), increase in respiratory rate and, in severe cases, pneumonitis or pulmonary oedema.

Irritancy (Sensory)

A response evoked in the eyes and upper respiratory tract by a toxicant, causing a painful sensation. This may be a direct stimulus of specialized receptors and secondary to tissue damage caused by the toxicant.

LC_{50}

Lethal concentration 50% - the concentration statistically calculated to cause the deaths of half the animals exposed to a toxicant for a specified time. It may be expressed in parts per million by volume (ppm), or milligrams per litre (mg/L). In combustion toxicology, two quite different values are often used: a) the LC_{50} expressed as mg/L which is based on the initial weight of the material divided by the volume of available air, and b) the LC_{50} expressed as mg/L which is based on the weight of the material actually consumed divided by the volume of available air. Care must be taken to distinguish between the two. As explained under dose, the time of exposure is very important in inhalation toxicology, and the duration of exposure should always be quoted for an LC_{50}.

LCT_{50}

The exposure dose statistically calculated to cause the death of 50% of the animals. This value is useful for comparing results obtained in experimental regimes employing different exposure times.

LT_{50}

The time of exposure statistically calculated to cause the death of 50% of the animals upon exposure to a fixed concentration of the toxicant.

Narcosis

Literally "sleep inducing", but used in combustion toxicology to describe central nervous system depression causing reduced awareness and impaired ability to escape. At higher concentrations of toxicants, unconsciousness and finally death will occur.

Pneumonitis

Inflammation of the lower respiratory tract.

Pulmonary Oedema

Extravasation of blood plasma in the alveolar regions of the lung caused by vascular damage, inflammation or inadequate venous drainage. The build up of fluid impairs absorption of oxygen into the blood.

RD_{50}

Respiratory depression 50% - a statistically calculated concentration of an irritant required to reduce the breathing rate of laboratory rodents by 50% during a 10- to 15-minute exposure.

Respiratory Tract

The nose, pharynx, larynx, trachea and large bronchi termed the upper respiratory tract, and the bronchioli, alveolar ducts and alveoli, termed the lower respiratory tract.

Specific Toxicity

A particular adverse effect caused by a toxicant, *e.g.*, narcosis, irritancy.

Supertoxicant

A term originally used to describe a toxicant with an unusual specific toxicity, often with a high toxic potency, not usually associated with fire effluents; use of the term is discouraged.

Toxic Hazard

A toxic hazard exists when a toxicant is present at a sufficient concentration and over a sufficient period of time to cause a toxic effect.

Toxic Potency

A measure of the dose of toxicant required to elicit a specific toxic effect - the smaller the dose required, the greater the potency.

Toxicant

A chemical capable of exerting an adverse effect or effects on an organism. The toxicant can be characterized by two properties: the specific toxicity, the nature of the adverse effect, and the toxic potency, the dose required to cause the effect.

Toxicity

The nature and extent of adverse effects of a substance upon a living organism.

REFERENCES

Alarie, Y. (1966). Irritating Properties of Airborne Material Respiratory Tract. *Arch. Environ. Health* **13**: 433-449.

Alarie, Y., Lin, C. K. and Geary, D. L. (1974). Sensory Irritation Evoked by Plastic Decomposition Products. *Am. Ind. Hyg. Assoc. J.* **35**: 654-661.

Amdur, M. O. (1957). The influence of Aerosols Upon the Respiratory Response of Guinea Pigs to Sulfur Dioxide. *Am. Ind. Hyg. Assoc. Quart.* **18**: 149-155.

Baetjer, A. M. and Smith, R. (1956). Effect of Environmental Temperature on Reaction of Mice to Parathion, an Anticholinesterase Agent. *Am. J. Physiol.* **186**: 39-46.

Burney, R. E., Wu, S. C., Nemiroff, M. J. (1982). Mass Carbon Monoxide Poisoning: Clinical Effects and Results of Treatment in 184 Victims. *Ann. Emerg. Med.* **11**: 394.

Cobb, L. M. (1981). Pulmonary Toxicity. **In**: *Testing for Toxicity* (J. W. Gorrod, Ed.). Taylor and Francis, pp. 255-274.

Durham, W. F., Wolfe, H. R. and Elliott, J. W. (1972). Absorption and Excretion of Parathion by Spraymen. *Arch. Environ. Health* **24**: 381-387.

Enna, S. J. and Schanger, L. S. (1960). Drug Absorption from the Lung. *Fed. Proc.* **28**: 359.

Fuhrman, G. and Fuhrman, F. A. (1961). Effects of Temperature on the Action of Drugs. *Ann. Rev. Pharmacol.* **1**: 65-78.

Funckes, A. J., Hayes, G. R., Jr. and Hartwell, W. V. (1963). Urinary Excretion of Paranitrophenol by volunteers Following Dermal Exposure to Parathion at Different Ambient Temperatures. *J. Agric. Food Chem.* **11**: 455-457.

Grigorowa, R. and Binnewies, S. (1973). Ueber die Kombinierte Wirkung von Phosphororganischen Pestiziden und Erhoehter Umgebungstemperatur Inhalatorischen Kurzuersuchen an Ratten: I. Toxikologische Aspekta. *Int. Arch. Arbeitsmed.* **31**: 295-307.

Kennedy, G. L. (1988). Techniques for Evaluating Hazards of Inhaled Products. **In**: *Product Safety Evaluation* (S. C. Gad, Ed.). Marcel Dekker, NY, pp. 259-289.

Keplinger, M. L., Lanier, G. E. and Deichmann, W. B. (1959). Effects of Environmental Temperature on the Acute Toxicity of a Number of Compounds in Rats. *Toxicol. Appl. Pharmacol.* **1**: 156-161.

Larkin, J. M., Brahos, G. J., Moylan, J. A. (1976). Treatment of Carbon monoxide Poisoning: Prognostic Factors. *J. Trauma* **16**: 111.

Lee, K. P. (1985). Lung Response to Particulates with Emphasis on Asbestos and Other Fibrous Dust. *CRC Crit. Rev. Toxicol.* **14**: 33.

Leong, B. K. J. (1981). *Inhalation Toxicology and Technology*. Ann Arbor Science.

Lu, F. C. (1985). *Basic Toxicology*. Hemisphere Publishing Corporation, New York.

McBay, A. J. (1965). Law Medicine Notes: Carbon Monoxide Poisoning. *N. Engl. J. Med.* **272**: 252.

Menzel, D. B. and McClellan, R. O. (1980). Toxic Responses of the Respiratory System. **In**: *Casarett and Doull's Science of Poison, 2nd Ed.* (J. Doull, C. D. Klaassen and M. O. Amdur, Eds.). Macmillan, New York.

Morrow, P. E. (1960). Some Physical and Physiological Factors Controlling the Fate of inhaled Substances. *Health Physics* **2**: 366-378.

Munson, E. S. and Eger, E. I. (1971). Pulmonary Disposition of Drugs. **In**: *Fundamentals of Drug Metabolism and Drug Disposition* (B. N. La Du, H. G. Mandel and E. L. Way, Eds.). Williams and Wilkins, Baltimore, 106-118.

Nelson, G. O. (1971). Controlled Test Atmospheres. Ann Arbor Science Publishers, Inc., Ann Arbor, MI.

Taylor, A. E., Guyion, A. C. and Bishop, V. B. (1965). Permeability of the Aveolar Membrane to Solutes. *Cir. Res.* **16**: 353-362.

Weihe, W. H. (1973). The Effect of Temperature on the Action of Drugs. *Ann. Rev. Pharmacol.* **13**: 409-425.

Weihe, W. H. (1976). The Effects on Animals of Changes in Ambient Temperature and Humidity. **In**: *Laboratory Animal Handbook, Control of the Animal House*. (T. McSheehy, Ed.). Laboratory Animals, Ltd., London, pp. 41-50.

Chapter 4

THE TOXICITY OF SMOKE AND COMBUSTION GASES

Shayne C. Gad

The atmosphere produced by the combustion of polymers is an exceedingly complex one. In the broadest sense (and as used generally in combustion toxicology), smoke is defined as a complex mixture of solid and liquid aerosols and gases produced when a material undergoes thermal decomposition (Kaplan *et al.*, 1983). In this chapter (and book) we will use the term in a much more restricted sense (airborne solid and liquid aerosols and the materials associated with their surface), because I believe that the use of the broader definition has caused the materials encompassed in the more narrow class (and their biological effects) to be ignored.

Thermal decomposition of a material may occur as a result of low oxygen (anaerobic) pyrolysis, oxidative pyrolysis (commonly referred to as "smoldering") and/or flaming combustion. Although all of these processes may occur at some point in a real fire, few fires start or progress in low oxygen atmospheres (NAS, 1978). Consequently, almost all laboratory scale test methods have been designed to decompose materials under oxidative pyrolysis and/or flaming combustion conditions.

When a material is thermally decomposed, a wide variety of chemical species are produced. Autian (1970) identified and quantified 50 chemical compounds detected in the volatile combustion products of a commercial sample of polyvinyl chloride; Mitera and Michal (1985) identified 55 compounds in the combustion products of polyethylene, 56 in the combustion products of polypropylene and more than 30 in those of both polystyrene and polyamide. Even with a natural material such as Douglas fir, Packham and Hartzell (1981) identified more than 75 different chemicals in the smoke produced by the combustion of this wood. The specific thermal decomposition products of any material, as well as the quantities and rates of evolution of these products, are highly dependent on the conditions under which the material is decomposed. The conditions include temperature, rate of heating, combustion mode (nonflaming or flaming) and oxygen availability (Clarke *et al.*, 1982; Kaplan *et al.*, 1983; Gad and Smith, 1983).

Almost all natural and synthetic polymers contain carbon and, when combusted, generate carbon monoxide (CO) and/or carbon dioxide (CO_2). The ratio of the quantities of the two gases evolved depends on the amount of oxygen present at the site of combustion. In addition to these two prevalent toxicants, various polymer decomposition products are released as simple saturated and unsaturated hydrocarbons (*e.g.*, methane, ethane, ethylene), partially oxidized species (*e.g.*, acetaldehyde, acrolein) and more complex aromatics (*e.g.*, benzene, toluene). Many materials contain nitrogen, sulfur and halogens in addition to carbon and hydrogen. When these materials are thermally decomposed, hydrogen cyanide (HCN), nitrogen oxides (NO_X), sulfur dioxide (SO_2), ammonia (NH_3) and halogen acids (HCl, HBr and HF) may be evolved. Also, other decomposition products, including isocyanates and nitriles may be formed. The common gases (other than CO and CO_2) produced by thermal decomposition of materials and the acute toxic effects of these gases are summarized in Table 4-1 (Terrill *et al.*, 1978). The toxic and physiologic effects of these gases will be examined in more detail later in this chapter.

The large number of diverse chemical species present in smoke must affect many organ systems of the body and cause myriad physiological alterations. Many of these alterations are masked by the effects of the major toxicants (CO, HCN and irritant gases) which are the most prevalent components of smoke and generally are present in the highest concentrations. It has thus become common practice in combustion toxicology to categorize the major fire effluents into two classes, the hypoxia-producing gases or asphyxiants and the irritants. A third large class has been designated for those chemicals with "other and unusual specific toxicities" (Hartzell *et al.*, 1985). Many components of smoke may fit into this category, but these compounds generally are present in low concentrations and are not analyzed for. With a few exceptions, the effects of these chemicals have not been evident in postmortem examinations of fire victims or in experimental studies with animals. Nevertheless, it is possible that some of these compounds contribute to the incapacitating and lethal effects of smoke inhalation, even though the effect of the hypoxia producing toxicants and the irritants predominate.

POSTMORTEM STUDIES OF FIRE VICTIMS

The search for the specific toxic agents responsible for inhalation fatalities has advanced along two lines. In the first, data have been obtained from the postmortem examinations of fire victims in an attempt to identify the causes of death. Some studies have analyzed data accumulated from many fires whereas others have investigated the fatalities in specific fires. The

Table 4-1. Toxicological Effects of Common Fire Gases

Toxicant	Sources	Toxicological Effects	Estimate of short-term (10 min.) lethal concentration (ppm)
Hydrogen cyanide (HCN)	From combustion of wool, silk, polyacrylonitrile, nylon, polyurethane, and paper.	A rapidly fatal asphyxiant poison.	350
Nitrogen dioxide (NO_2) and other oxides of nitrogen (NO_x)	Produced in small quantities from fabrics and in larger quantities from cellulose nitrate and celluloid.	Strong pulmonary irritant capable of causing immediate death as well as delayed injury	> 200
Ammonia (NH_3)	Produced in combustion of wool, silk, nylon, and melamine, concentrations generally low in ordinary building fires.	Pungent, unbearable odor; irritant to eyes and nose.	> 1000
Hydrogen chloride (HCL)	From combustion of polyvinyl chloride (PVC), and some fire-retardant-treated materials.	Respiratory irritant; potential toxicity of HCl absorbed on surface of particulates is absent particulates is greater than that for an equivalent amount of gaseous HCl.	> 500, if particulate is absent
Other halogen acid gases (HF, COF_2 and HBr)	From combustion of fluorinated resins or films and some fire-retardant materials containing bromine.	Respiratory irritants acting much as HCL. Toxicity increased by presence of fine particulates.	HF ? 400 HBR > 500
Sulfur Dioxide (SO_2)	From materials containing sulfur.	A strong irritant, intolerable well below lethal concentrations.	> 500
Isocyanates	From urethane polymers; pyroly-products, such as toluene-2,4-diisocyanate (TDI), have been reported in small scale laboratory studies; their significance in actual fires is undefined.	Potent respiratory irritants; believed the major irritants in smoke of isocyanate-based urethanes.	~ 100 (TDI)
Acrolein	From pyrolysis of polyolefins and cellulosics at lower temperatures (400°C)	Potent respiratory irritant.	30 to 100

Source: Modified from Terrill *et al.*, 1978

Table 4-2. Distribution of Fire victims According to Blood CO Saturation Levels

COHb (%)	Number of Victims	Percent
0-9	48	9
10 - 19	42	8
20 - 29	37	7
30 - 39	38	7
40 - 49	43	8
50 - 59	58	11
60 - 69	79	15
70 - 79	111	21
>80	74	14
Total	530	100

Source: Birky *et al.*, (1979).

results of these studies have established that carbon monoxide is the primary cause of death of most fire victims, and that other factors, including cardiovascular disease, alcohol, HCN, HCl and particulates, have been or may have been involved in some deaths.

A major study of postmortem data was a 6-year study by Johns Hopkins University of 530 fire fatalities from 398, primarily residential, fires in Maryland (Birky *et al.*, 1979)(Table 4-2). The postmortem examinations consisted of 1) analyses of carboxyhemoglobin, cyanide and alcohol in the blood and other tissues, 2) heavy metal analyses of trachea and bronchi, and 3) pathological examinations of victims hearts. Results of analyses for blood COHb showed that approximately 60 percent of the victims had blood COHb levels equal to or greater than 50 percent, which is generally considered evidence of CO-induced death (Halkin and Berl, 1979). In 40 percent of the victims in which COHb levels were below 50 percent, the investigators attributed death to CO plus cardiovascular disease in 20 percent, to burns in 11 percent, and to "unexplained" causes (possibly other toxicants) in 9 percent of the victims. In a similar 3-year study of 182 fire fatalities in Glasgow, Scotland, lethal blood levels of COHb were measured in most of the victims and heart disease was prevalent in victims in which COHb levels were below 50 percent (Clarke, 1982). In both the Maryland and Glasgow studies, intoxicating levels of alcohol were present in a large percentage of the victims.

Hydrogen cyanide has been implicated as a possible cause of smoke-inhalation fatalities by a number of studies which have shown cyanide in the blood of victims at levels considered toxic or even lethal. In the 272 Maryland fire fatalities in which blood cyanide concentrations were measured, almost 70 percent had levels above "normal" (0.25 μg/mL) and 10 percent had levels considered "probably toxic" (> 2.00 μg/mL) by the investigators. In another study, in which the blood of ten of 42 victims of the Tennessee jail fire was analyzed for COHb and cyanide, all of the victims except one had elevated cyanide levels (Birky *et al.*, 1979). However, elevated blood cyanide levels in victims of this and other fires were almost always accompanied by lethal levels of COHb. One exception was the Houston Westchase Hilton fire in which blood cyanide levels were elevated in nine of ten victims but COHb levels of 50 percent or greater were present in only two victims (Birky, 1982). The cause of death of two victims in which 8.0 and 10.0 μg/mL of cyanide were measured was attributed by the Medical Examiner to "asphyxia due to soot, carbon monoxide and cyanide inhalation." Because high COHb levels almost always accompany elevated cyanide levels, it has not been possible to determine the singular role of HCN in causing fatalities attributed to smoke inhalation. In addition, blood cyanide data appear to be less definitive than COHb levels because of the difficulties involved in analyzing blood cyanide, possible blood storage effects on cyanide concentrations and incomplete understanding of the pharmacokinetics of HCN (Ballantyne, 1976). It is the opinion of some investigators that elevated cyanide and lethal COHb levels in the blood of fire victims indicate that HCN caused early incapacitation of the victims, resulting in continued inhalation of CO and subsequent death (Birky, 1982; Pursler and Wooley, 1983).

Irritant gases are also prevalent in combustion atmospheres and are capable of producing severe effects on the eyes and respiratory tract. However, the contribution of these gases to fire fatalities that have been attributed to smoke inhalation is not well understood. Evidence that these gases may play a significant role in some deaths has been based primarily on findings of respiratory tract damage in some victims, because it is rarely possible to detect irritant gases in postmortem analyses of blood and othertissues. In the Westchase Hilton Hotel fire, hydrogen chloride was considered to have had a contributory role in some of the fire fatalities (Birky, 1979). The basis for this conclusion was the detection of chloride in soot samples from the room of origin and pathological findings of respiratory tract damage and pulmonary edema in two fatalities and of corneal burns in one survivor of the fire. In an earlier report, Dyer and Esch found extensive pulmonary hemorrhaging and edema in a fireman who died 24 hours after being present in a fire involving PVC Teflon (Dyer and Esch, 1976).

TOXIC EFFECTS OF COMMON COMBUSTION ATMOSPHERE GASES

As can be imagined from the previous section, the essential element in evaluating the toxicological hazards of combustion atmospheres is the determination of the major toxicants responsible for the observed toxic effects. Analytical data alone are generally insufficient for this determination because gases which are not routinely analyzed for may be the significant toxicants. In addition, chemical analyses cannot detect those toxic effects that may occur as a result of biological interactions between toxicants (such as chloramines). Consequently, any analytical data must be correlated with toxicological results to determine whether the measured toxicants are the likely causative species or whether some unanalyzed component has a significant influence. The determination of whether common toxicants are responsible for the observed toxic effects is based on toxicity data obtained in studies with the pure gases.

CARBON MONOXIDE

There is an extensive database on the incapacitating and lethal effects of CO in rats to help determine whether the observed toxicity of a combustion atmosphere is due solely to CO intoxication (Kaplan and Hartzell, 1984). In a study of CO by Hartzell and others (1985), concentration-time curves for incapacitation using leg-flexion shock avoidance and for lethality (Figure 4-1, adapted from Kaplan, 1988) were linearized by plotting the concentration of CO against the reciprocal of the time to incapacitation or to death. At CO concentrations of 2000 to 10,000 ppm, the C•t products which incapacitate the rat range from approximately 38,000 to 40,000 ppm-minutes. The C•t products necessary to cause lethality in the rat range from 125,000 to 147,000 ppm-minutes at concentrations of 5,000 to 10,000 ppm and are approximately three times greater than those necessary to incapacitate the rat. As a guide, a C•t product of approximately 35,000 ppm-min or greater may be considered evidence that CO was responsible for incapacitation of the rat by a combustion atmosphere and a C•t product of approximately 100,000 ppm-min is evidence of CO-induced lethality. The blood COHb saturation levels in rats incapacitated by CO generally are 65 percent or greater and the levels in animals that die from exposure to CO generally exceed 75 percent (Kaplan, 1988).

Carbon monoxide is also directly cardiotoxic, but has been shown to not account quantitatively for the total cardiotoxicity of combustion atmospheres (Thomas and O'Flaherty, 1982).

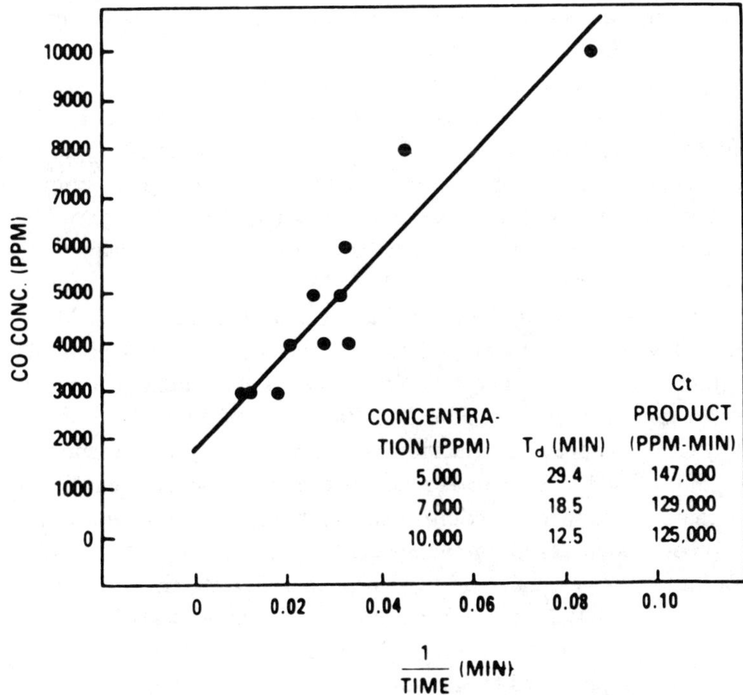

Figure 4-1 Concentration-time curve for lethality of rats by CO and Ct products associated with lethality. (Adapted from Kaplan, 1988)

CARBON MONOXIDE AND SENSORY IRRITANTS

Data from combustion toxicity tests of selected materials have established that incapacitation and, possibly, lethality in rats occurs later than expected from the concentration-time curves for CO. Although incapacitation or death appeared to have been delayed, blood COHb saturation levels were consistent with a CO-induced effect, *i.e.*, > 65 percent COHb at incapacitation and > 75 percent COHb at death. In such tests there generally are clinical signs of the presence of sensory irritants, including severe irritation of the eyes, respiratory abnormalities, severe weight loss and post exposure lethalities. Because sensory irritants cause a reflex inhibition of respiratory rate in rodents, these delayed effects of CO could be due to a decreased rate of loading of CO. This inhibition of respiration by irritants in rats is not observed in nonhuman primates which respond to high concentrations of irritant gases with a brief and transient holding of the breath followed by an increase in respiratory rate (Kaplan *et al.*, 1986).

The effect of HCl on CO intoxication in rats has been investigated by Hartzell *et al.* (1985). These investigators reported that HCl caused decreases in respiratory rate, ranging from a 35 percent decrease at 200 ppm to a 67 percent decrease at 1500 ppm. Where rats were exposed to 1000 ppm of HCl in combination with various concentrations of CO, CO (3000 to 5000 ppm) C•t products were approximately 65,000 ppm-minutes for incapacitation or almost 75 percent greater than the C•t products necessary for incapacitation by CO alone. Thus, times to incapacitation from exposure to 3000 to 5000 ppm of CO were almost doubled in the presence of 1000 ppm of HCl. Although the data were highly variable 2500 ppm HCl did not cause a greater increase in time to incapacitation than 1000 ppm of HCl. Time to lethality caused by CO also appeared to be increased somewhat by the presence of 1000 ppm or 2500 ppm HCl, but not as much as time to incapacitation. These data indicate that, in combustion toxicity tests of materials, the presence of irritants may "protect" rodents from the incapacitating or lethal effects of CO during the 30-minute exposure and lead to an artificial increase in the IC_{50} or LC_{50} value of the material. This is one reason why screening tests based on toxic potency of combustion atmospheres in rodents may yield invalid data for evaluating the toxic hazard to humans.

HYDROGEN CYANIDE

Although the incapacitating and lethal effects of HCN have not been studied as extensively as those of CO, there is a sufficient data base to evaluate whether HCN is a major toxic component of a combustion atmosphere. It is evident from the linearized concentration-time curves for incapacitation (Figure 4-2) that even low concentrations of HCN are rapidly incapacitating and that Haber's rule is valid for a much narrower range of concentrations of HCN than of CO. Even so, at HCN concentrations of approximately 150 to 500 ppm, a C•t product of approximately 800 ppm-minutes or greater may be considered evidence that HCN was responsible for the incapacitation of test animals. Thus, on the basis of C•t values, the intensity of exposure required to incapacitate rats with HCN is approximately 2% of the intensity required with CO, except perhaps at the extreme ends of the time-concentration curves, where large deviations from Haber's rule occur. The concentration-time relationship for lethality (Figure 4-2) also shows greater deviation from Haber's rule with HCN than with CO. These data, as well as data from studies by Levin *et al.* (1986) and by Higgins *et al.* (1972), indicate that a C•t product of approximately 2500 ppm-minutes or

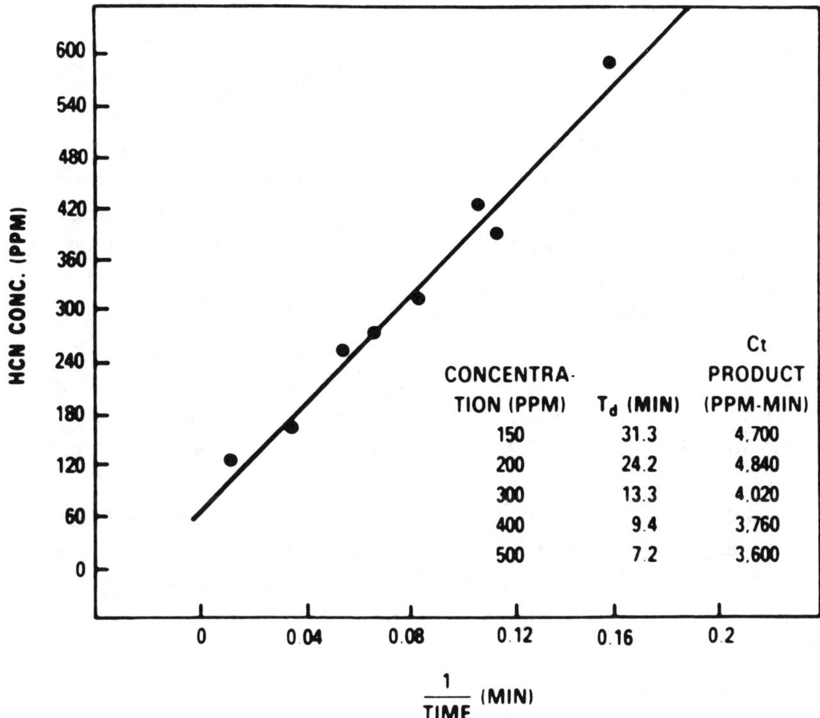

Figure 4-2 Concentration-time curve for lethality of rats by HCN and Ct products associated with lethality. (Adapted from Kaplan, 1988)

greater may be used as a guide to evaluate whether HCN was responsible for the deaths of animals in combustion toxicity tests. This product is approximately one-fortieth the C•t product necessary for CO to cause lethality.

HYDROGEN CYANIDE AND CARBON MONOXIDE

Because CO interferes with the transport of oxygen by the blood and HCN interferes with the utilization of oxygen by cells, one might expect that exposure to the two gases simultaneously would result in additive effects. Experimental studies of the combined effects of HCN and CO, however, have yielded conflicting results. In a study by Higgins and others (Higgins *et al.*, 1972) LC_{50} values of a 5-minute exposure to HCN in rats (503 ppm) and in mice (323 ppm) were not statistically different from those of a mixture of HCN and CO at a concentration that produced a 25-percent blood COHb level. Even when animals were exposed to HCN after an exposure to CO that produced blood COHb levels of 25 or 50 percent, additive effects of the two gases were not evident. In contrast, Smith (1976) and Lynch (1975) have reported that, based on times to incapacitation and death and $L(C{\bullet}t)_{50}$ values, respectively, simultaneous exposure to HCN and CO results in additive effects. The results of incomplete experiments at Southwest Research Institute (Kaplan, 1988) also suggest that the two gases are additive when time to incapacitation or death is measured, but low or high concentrations of either gas may yield inconsistent and considerably variable results. In view of the inconsistent results obtained by different investigators, a comprehensive study of the combined effects of CO and HCN is warranted. Until more definitive data are available, it should be assumed that additive effects do occur and a combined C•t product should be calculated for the two gases to evaluate whether HCN and CO were responsible for incapacitation or death of animals. A combined C•t product for CO equivalents may be calculated as the product of the exposure time and the sum of the CO concentration and 40 times (the equivalent potency factor) the HCN concentration. Accordingly, a mixture of 200 ppm of HCN and 4000 ppm CO would have a combined equivalent concentration of 8000 ppm of CO and would be expected to incapacitate rats in approximately 4 to 5 minutes (35,000 ppm-min CO equivalents).

IRRITANT GASES

In combustion toxicity tests of materials, the toxic effects of irritant gases generally consist of irritation and lacrymation of the eyes (even corneal opacities in extreme cases), respiratory abnormalities, body weight loss and post-exposure lethalities. The severity of these effects varies with the specific irritants present in the smoke and their concentrations. Gross pathological and histopathological abnormalities may be evident at necropsy in the eyes and respiratory tract of animals exposed to irritant gases. Although the NBS test method protocol provides for the observation of clinical signs and the pathological examination of animals, most combustion toxicity tests have neglected the toxic effects of irritant gases, except for the measurement of post-exposure lethality. The notable exception is the current University of Pittsburgh test protocol which evaluates sensory irritation by measurement of respiratory rate depression in mice and provides for determination of RD_{50} values of gases and combustion atmospheres. It is apparent from the RD_{50} values reported by Barrow *et al.* (1978) in Table 4-3 that gases as well as combustion atmospheres produced by materials may differ considerably in the severity of their irritating effects and, consequently, in their potential hazard to humans in a fire environment.

Table 4-3. RD_{50} Concentrations for the Thermal Decomposition Products of Various Polymers and Several Single Airborne Chemical irritants

Thermal decomposition products for various polymers limits	RD_{50} (mg/L)	95% Confidence
Flexible polyurethane foam	0.06	0.04-0.09
Fiberglass-reinforced polyester	0.14	0.12-0.17
Polystyrene (oxidative pyrolysis)	0.18	0.08-0.42
Douglas fir	0.24	0.15-0.35
Polytetrafluoroethylene (no plasticizer or fire retardant)	0.25	0.18-0.35
Polyvinyl chloride (no plasticizer or fire retardant)	0.50	0.32-0.78
Polyvinyl chloride (plasticized)	0.19	0.17-0.22
Polystyrene (flaming combustion)	6.0	4.4-8.4

Single chemicals	RD_{50} (ppm)	95% Confidence limits	RD_{50} (mg/L)
Toluene diisocyanate	0.39	0.31-0.49	0.0028
Chlorine	9.3	5.2-16.9	0.027
Sulfur dioxide	117	67-204	0.31
Ammonia	303	188-490	0.21
Hydrogen chloride	309	219-435	0.45

Source: Adapted from Barrow *et al.*, 1978

The identification of irritant gases as major toxicants in combustion atmospheres is much more difficult than the identification of either CO or HCN, for several reasons. The first reason is that, except for respiratory rate depression, the only effect of irritants that is measured in combustion toxicity tests and can be correlated with analytical data is post-exposure lethality. This is a delayed effect, however, to which other toxicants and factors, including animal health status and environmental conditions, may contribute. Consequently, there may be considerable variability in the occurrence of post-exposure lethality in animals exposed to an irritant gas or a combustion atmosphere. A second reason is that since studies of relatively few pure irritant gases have been conducted, there is a limited data base for identifying an irritant gas as a major toxicant. Finally, with a few exceptions, automated instrumentation for analyses of irritant gases is not available, and the analyses of these gases generally are more complex and time-consuming than that of CO or HCN.

Hydrogen chloride has received the most attention in studies of the irritant gases. Most of the experimental work with HCl has been conducted with rodents, although a few studies with nonhuman primates have been reported. The rat can tolerate high concentrations of this gas without incapacitation or lethality occurring during exposure. Sufficiently high concentrations (greater than approximately 80,000 ppm) have been reported to incapacitate the rat in the motor-driven exercise wheel and shuttlebox within a few minutes, with death occurring shortly thereafter (Kaplan *et al.*, 1985; Crane *et al.*, 1985). A number of investigators have reported LC_{50} values in rats exposed to HCl for different time periods. There are discrepancies between these values, which may be not significantly different than those obtained in rats exposed to each gas with CO sufficient to produce a 25-percent COHb level. These may be due to differences in the post-exposure periods of observation as well as to difficulties in analyzing HCl. For example, Darmer *et al.* (1974) reported LC_{50} values of 40,989 ppm (34,803-48,272) for a 5-minute exposure and of 4701 ppm (4129-5352) for a 30-minute exposure, with 7-day post-exposure observation periods. Hartzell *et al.* (1985) reported an LC_{50} value of 15,900 ppm (11,540-21,890) for a 5-minute exposure of rats and an LC_{50} value of 3,715 ppm (2,540-5,435) for a 30-minute exposure, but lethalities during the 14-day post-exposure period were included in these values. These investigators also determined LC_{50} values for other exposure times. From a linearized plot of the LC_{50} value as a function of the reciprocal of exposure time (Figure 4-3) it is evident that the range of $L(C\bullet t)_{50}$ products is wide, varying from 80,000 ppm-minutes at a concentration of 16,000 ppm to 132,000 ppm-minutes at 4,000 ppm of HCl. The data reported in this study indicate that a C•t product of approximately 50,000 ppm-min may be used as a guide to evaluate whether HCl in a combustion atmosphere may have been responsible for post-exposure deaths.

Limited data are available for assessing whether other irritants are major toxicants in the combustion atmospheres of materials. In an unpublished study of acrolein by Crane (Kaplan, 1988), rats were incapacitated in the motor-driven exercise wheel after a 5-minute exposure to from 5,000 to 10,000 ppm of acrolein, with death occurring shortly thereafter. Higgins *et al.* (1972) investigated the toxicity of 5-minute exposures of mice and rats to HF and NO_2, with and without CO. The LC_{50} values (7-day post-exposure observations) in rats of 18,200 ppm (15,965 20,748) for HF and 831 ppm (566-1,240) for NO_2 were not significantly different from those obtained in rats exposed to each gas with CO sufficient to produce a 25% COHb level.

CARBON DIOXIDE

Carbon dioxide stimulates the respiratory center and is capable of causing a marked increase in the respiratory minute volume of animals. Ventilation in humans is increased 50% by a concentration of 18,000 ppm of CO_2 in air and 100% by a concentration of 25,000 ppm (Kimmerle, 1974). This increased

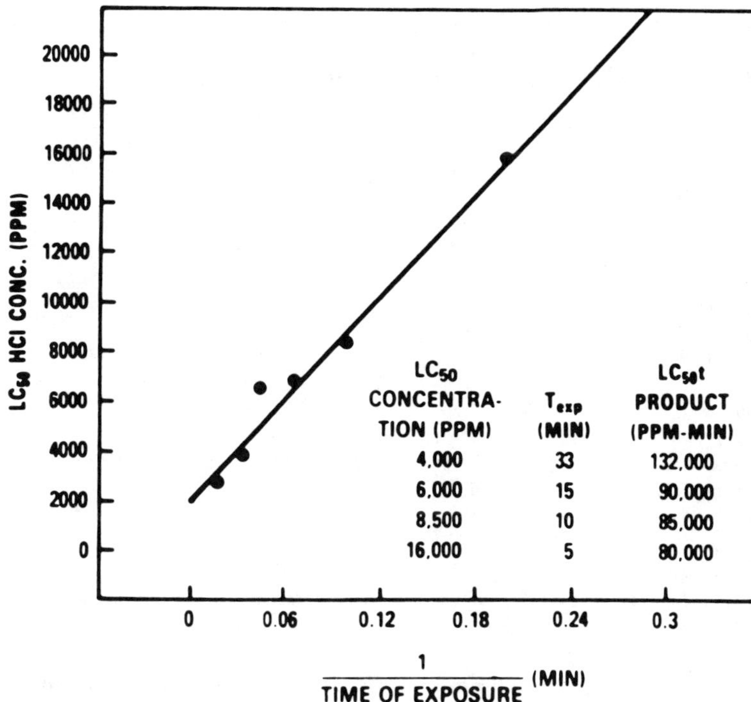

Figure 4-3 LC_{50} values of HCl in rats (14-day postexposure observation) as a function of exposure time and Ct products for the LC_{50} values. (Adapted from Kaplan, 1988)

ventilatory effect of CO_2 would be expected to increase the uptake of CO and other gases and possibly cause an increase in their toxicity. Studies of the combined effects of CO and CO_2, however, have yielded conflicting results. Levin found that 2500 ppm of CO is twice as toxic in the presence of 5% CO_2, with lethality as the index of toxicity. In a study by Rodkey and Collison (1979), the presence of 4.5% CO_2 was found to decrease the time of death of rats exposed to 6000 ppm of CO from 22.4 ± 0.8 min to 16.8 ± 0.6 min. In contrast to these results, Crane (1985) did not find a significant change in the time to incapacitation or death of rats exposed to 5000 ppm of CO when concentrations of 4 to 13% of CO_2 were added to the CO. Similarly, Hartzell and Switzer (1985) reported that 5% CO_2 did not significantly change the time to incapacitation of rats exposed to 3500 ppm of CO.

Table 4-4. Hazardous Concentrations of Major Toxicants Generated by Thermal Decomposition of Materials

Toxicant	ACGIH TLV-C[a] (ppm)	ACGIH TLV-STELb (ppm)	IDLH[c] (ppm)	STLC[d] (ppm)	Hazardous/Lethal within a few mins[e] (ppm or %)
Carbon monoxide	—	400	500	5000	8000
Hydrogen cyanide	10	—	50	350	280
Carbon dioxide	—	30,000*	50,000	—	12%
Oxygen depletion	—	—	—	—	<12%
Hydrogen chloride	5	—	100	>500	1000-2000
Hydrogen bromide	3*	—	50	>500	—
Hydrogen fluoride	3*	—	20	400	50-100
Ammonia	—	35	50	>1000	2000
Acrolein	—	0.3	5	30-100	>10
Formaldehyde	—	2*	100	—	50
Sulfur dioxide	—	— **	100	>500	600-800
Nitrogen dioxide	—	5	50	>200	250
Styrene	—	100	5000	—	—
Toluene-2,4-diisocyanate	—	0.02	10	100	—

[a]. American Conference of Governmental Industrial Hygienist (1984-85) Threshold Limit value - Ceiling - The concentration that should not be exceeded even instantly in the work environment.

[b]. ACGIH Threshold Limit Value - Short-Term Exposure Limit - The concentration to which workers can be exposed continuously for a short period of time (15 minutes) without suffering from 1) irritation, 2) chronic or irreversible tissue damage, or 3) narcosis of sufficient degree to increase the likelihood of accidental injury, impair self-rescue or materially reduce work efficiency.

[c]. Immediately Dangerous to Life or Health - The maximum concentration from which one could escape within 30 minutes without any escape impairing symptoms or any irreversible health effects. These IDLH levels were established in the NIOSH/OSHA Standards Completion Program.

[d]. Short-Term (10-minutes) Lethal Concentrations from Terrill and others (1978).

[e]. From Kimmerele (1974) except for hydrogen chloride values from Haggard, H. W. and Y. Henderson, *Noxious Gases, 2nd Edition,* NY Reinhold, p. 126-127 (1943).

* Proposed change value.

** Proposed deletion of STEL value.

Table 4-4 summarizes currently accepted standards for what constitutes the lower limits of the major individual gases involved in combustion. Largely unaddressed, however, is what the effects of multicomponent mixtures of these gases (and of other, more exotic and reactive combustion atmosphere components) may be. Though some (Tsuchiya, 1986) have argued against there being any synergism operative in the biological effects of a combustion gas atmosphere, the data already reviewed in this chapter make clear that complex interactions are present.

SMOKE AND ITS TOXICITY

As pointed out at the beginning of this chapter, though most people characterize all the products of combustion as smoke, a rigorous definition is quite different. Smokes include both solid and liquid particulates, the latter usually being the aerosols formed by condensation of supersaturated vapors. These range in size (aerodynamic diameter) from 0.01 to 1.0 micron, and for our purposes also include the materials absorbed onto the surface of these aerosols. Smoke, by this definition, is a critical (but largely ignored) part of the combustion toxicity hazard problem.

Smoke is also an important fire-response characteristic because visibility is a factor in the ability of occupants to escape from a burning structure, and in the ability of firefighters to locate and suppress a fire.

All fires produce smoke, which may be thick or thin, light or dark. Any appreciable quantity of smoke invariably is irritating and obscures light and vision. A number of methods for measuring smoke obscuration have been suggested or used during the last few years. No universal, or even general, standard exists at present, but preference has been shown in the United States for the NBS Smoke Chamber Test (ASTM E 662). Flame retardants increase carbon in the pyrolysis residue and decrease the yield of volatile aromatic products.

Smoke in the early stages of a fire does not contain a high enough concentration of dangerous materials to be acutely lethal. However, the initially generated smoke, by its irritant properties and obscuration of normal visibility, immobilizes the occupants within the area of the building where they are located. They are then "trapped" within the building, and unless rescued promptly, may be killed by lethal heat and gases that follow. Smoke is the primary escape problem in building fires, in spite of the fact that heat, CO and gases are the major short term lethal components in the fire environment. Recent large-scale enclosure and occupancy tests suggest that heat, not CO or other toxicants, is the most likely second survivability criterion to be reached, usually 10 minutes after smoke bars the escape route.

Observations of the physical aspects of the smoke forming characteristics of various polymers include the following:

- Structures containing the aliphatic carbon backbone, such as polyolefins, nylons, acetal and PMMA, are low in smoke density. Most of these polymers melt and burn away in 1-2 minutes. PMMA is very low in smoke density, melting in about 1 minute. These resins are generally not self-extinguishing, and additives to increase the oxygen index usually increase smoke.

- Polymers containing aromatic rings, *e.g.* polystyrene, NORYL, SAN, and ABS have high smoke densities.
- Resins of high thermal stability, such as polysufone, polycarbonate, PPO, PET, *etc.* are intermediate in smoke density, polysulfone, in general, having the lower values.
- Alloys or blends, such as Arylon and Kydene (acrylic/PVC) show smoke density behavior intermediate between those of the base resins, but shifted slightly toward higher optical densities. The generation of smoke from some of the aromatic polymers is so voluminous that smaller standard samples or a larger test compartment would be preferred for accurately measuring smoke generation under laboratory conditions.
- Although polyethylene (PE) is very low in smoke density, substitution of a chlorine for a hydrogen atom increases smoke density to a very high degree (compare PE with PVC).
- In the case of polyolefins, as the melting point (and number of carbons) increases (from LDPE to poly-4-methylpentene), the smoke density increases. This observation may reflect the longer time in the flame for the higher melting resins before melting and dripping through the screen.
- Chlorinating PE results in very high smoke densities, compared to chlorinating PVC. On the other hand, substituting a chlorine for hydrogen in polyethylene (*i.e.* PVC) effects a substantial increase in smoke density.
- Ethylene-ethyl acrylate and ethylene vinyl acetate copolymers have low smoke densities and melt through the screen in about 1½ minutes.

Smoke particles also provide a vehicle for deep lung penetration of highly reactive molecular entities. These tend to cause damage which is not immediately lethal, but which nevertheless is extremely serious (Alexeeff *et al.*, 1986). Such a mechanism may be involved in the case of those materials that are associated with seemingly unexplainable delayed deaths.

REFERENCES

Alexeeff, G. V., Lee, Y. C., Thorning, D. Howard, M. L. and Hudson, L. D. (1986) Pulmonary Tissue Reactions in Response to Smoke injury. *J. Fire Sci.* **4**: 427-442.

Autian, J. (1970) Toxicologic Aspects of Flamability and Combustion of Polymeric Materials. *J. Fire Flam.* **1**: 239.

Ballantyne, B. (1976) Changes in Blood Cyanide as a Function of Storage Time and Temperature. *J. Forensic Sci. Soc.* **10**: 305.

Barrow, C. S., Alarie, Y., and Stock, M. F. (1978) Sensory Irritation and Incapacitation Evoked by Thermal Decomposition Products of Polymers and Comparisons with Known Sensory Irritants. *Arch. Environ. Health* **33**: 79.

Birky, M. M. (1982) *Wetchase Hilton fire report, Part II*. The Foundation for Fire Safety News, Rosslyn, Virginia, pp. 1-17.

Birky, M. M., Halpin, B. M., Caplan, Y. H., Fisher, R. S., McAllister J. M. and Dixon, A. M. (1979) Fire fatality study. *Fire and Materials* **3**: 211 (1979).

Birky, M. M., Paabo, M. and Brown, J. E. (1979) Correlation of Autopsy Data and Materials Involved in the Tennessee Jail Fire. *Fire Safety J.* **2**: 17.

Clarke, F. B., III, Benjamin, I. A. and Clayton, J. W. (1982) *An Analysis of Current Knowledge in Toxicity of the Products of Combustion.* Report to Committee on the Toxicity of the Products of Combustion of the National Fire Protection Association, Morgan Technical Library, Batterymarch Park, Quincy, Massachusetts.

Crane, C.R. (1985), Are the combined toxicities of CO and CO_2 Synergistic (?) *J. Fire Sci.* **3**: 143.

Crane, C. R., Sanders, D. C., Endecott, B. R. and Abbott, J. K. (1985) *Inhalation Toxicology IV. Times to Incapacitation and Death for Rats Exposed Continuously to Atmospheric Hydrogen Chloride Gas* FAA-AM-85-4, Civil Aeromedical Institute, Federal Aviation Administration, Oklahoma City, Oklahoma.

Darmer, K. I., Jr., Kinkead, E. R. and DiPasquale, L. C. (1974) Acute Toxicity in Rats and Mice Exposed to Hydrogen Chloride Gas and Aerosols. *J. Am. I . Assoc.* **35**: 623.

Dyer, R. F., and Esch, V. H. (1976) Polyvinyl Chloride Toxicity in Fires, Hydrogen Chloride Toxicity in Fire Fighters. *J.A.M.A* **235**: 393.

Gad, S. C. and Smith, A. C. (1983) Influence of Heating Rates on the Toxicity of Evolved Combustion Products: Results and a System for Research. *J. Fire Sci.* **1**: 465.

Halpin, B. M. and Berl, W. G. (1979). Human Fatalities from Unwanted Fires. *Fire J.* **73**: 105.

Hartzell, G. E., Priest, D. N. and Switzer, W. G. (1985) Modeling of Toxicological Effects of Fire Gases: II. Mathematical Modeling of Intoxication of Rats by Carbon Monoxide and Hydrogen Cyanide. *J. Fire Sci.* **3**: 115.

Hartzel, G. E., Stacy, H. W., Switzer, W. G., Priest, D. N. and Packham, S. C. (1985) Modeling of Toxicological Effects of Fire Gases: IV. Intoxication of Rats by Carbon Monoxide in the Presence of an Irritatant. *J. Fire Sci.* **3**: 263.

Hartzell, G. E. and Switzer, W. G. (1985) On the Toxicities of Atmospheres Containing Both CO and CO_2. *J. Fire Sci.* **3**: 307.

Higgins, E. A., Fiorca, V., Thomas, A. A. and Davis, H. V. (1972) Acute Toxicity of Brief Exposure to HF, HCL, NO_2 and HCN with and without CO. *Fire Technology* **8**: 120.

Kaplan, H. L. (1988) Evaluating the Biological Hazards of Combustion Products. **In**: *Product Safety Evaluation Handbook* (S. C. Gad, Ed.) Marcel Dekker, N.Y., pp. 409- 7.

Kaplan, H. L., Anzueto, A. Switzer, W. G. and Hinderer, R. K. (1986) Respiratory Effects of Hydrogen Chloride in the Baboon. *The Toxicologist* **6**: 52.

Kaplan, H. L. Grand, A. F., Hartzell, G. E. (1983) *Combustion Toxicology: Principles and Test Methods.* Technomic Publishing Company Inc., Lancaster, PA, pp. 1-174 (1983).

Kaplan, H. L., Grand, A. F., Switzer, W. G., Mitchell, D. S., Rogers, W. R. and Hartzell, G. E. (1985) Effects of Combustion Gases on Escape Performance of the Baboon and the Rat. *J. Fire Sci.* **3**: 228.

Kaplan, H. L. and Hartzell, G. E. (1984) Modeling of Toxicological Effects of Fire Gases: I. Incapacitating Effects of Narcotic Fire Gases. *J. Fire Sci.* **2**: 286.

Kimmerele, G. (1974) Aspects and Methodology for the Evaluation of Toxicological Parameters During Fire Exposure. *J. Fire and Flamability/Comb. Tox.* **1**: 4.

Levin, B. C. (1985) Chemical and Engineering News, American Chemical Society, pp. 3 -3 , (May 20, 1985).

Levin, B. C., Gurman, J. L., Paabo, M., Baier, L., Procell, L., and Newball, H. H. (1986) Acute Inhalation Toxicity of Hydrogen Cyanide. *The Toxicologist* **6**: 59.

Lynch, R. D. (1975) *On the Non-existence of sYnergism Between Inhaled Hydrogen Cyanide and Carbon Monoxide.* Fire Research Note No. 1035, Fire Research Station, Borehamwood, England.

Mitera, J. and Michal, J. (1985) The Combustion Products of Polymeric Materials, II GC-MS Analysis of the Combustion Products of Polyethylene, Polypropylene, Polystyrene and Polyamide. *Fire and Materials* **9**: 11.

NAS (1978) *Smoke and Toxicity (Combustion Toxicology of Polymers), Vol. 3.* A Report by National Materials Advisory Board, Publication No. NMAB 318-3, National Academy of Sciences, Washington, D.C., pp. 1-55.

Packham, S. C. and Hartzell, G. E. (1981) Fundamentals of Combustion Toxicology in Fire Hazard Assessment. *J. Test, Eval.* **9**: 341.

Purser, D. A., Woolley, W. D. (1983) Biological Studies of Combustion Atmospheres. *J. Fire Sci.* **1**: 118.

Rodkey, F. L., Collison, H. A. (1979) Effects of Oxygen and Carbon Dioxide on Carbon Monoxide Toxicity. *J. Comb. Tox.* **6**: 208.

Smith, P. W. (1976) FAA Studies of the Toxicity of Products of Combustion. *Proceedings of the First Conference and Workshop on Fire Casualties.* Publication APL/JHU FPP 1-376-1, Johns Hopkins University, Baltimore, Maryland, pp. 173-184.

Terrill, J. B., Montgomery and R. R., Reinhardt, C. F. (1978) Toxic gases from fires. *Science* **200**: 1343.

Thomas, W. C., O'Flaherty, F. J. (1982) The Cardiotoxity of Carbon Monoxide as a Component of Polymer Pyrolysis Smokes. *Toxicol. Appl. Pharmacol* **63**: 363-372.

Tsuchiya, Y. (1986) On the Unproved Synergism of the Inhalation Toxicity of Fire Gas. *J. Fire Sciences* **4**: 346-354.

Chapter 5

COMBUSTION TOXICITY TESTING

Shayne C. Gad

Laboratory scale test systems are the traditional approach to evaluating the potential combustion toxicity of materials. These test system serve to generate combustion product streams from a sample and expose test animals to the stream. The response of test animals over a short period of time (from one day to two weeks) evoked by this acute exposure is then evaluated.

Laboratory scale test methods for evaluating the acute toxicity of combustion atmospheres generated by materials have been reviewed by Birky (1977), the National Academy of Sciences (1978), Kaplan *et al.* (1983), Kaplan (1987) and by Alarie (1985). In the review by Kaplan *et al.* (1983), the apparatus and methodology of the six most commonly utilized test system are described in detail, data obtained with each method are presented and the advantages and limitations of each method discussed. These methods are (1) the DIN method; (2) the Federal Aviation Administration (FAA) method; (3) the National Bureau of Standards (NBS) method; (4) the Radiant Heat Test method; (5) the University of Pittsburgh (PITT) method; and (6) the University of San Francisco (USF method). In addition to these, several additional methods which have been used in the combustion toxicity testing of materials were reviewed. This chapter seeks to update and supplement these earlier presentations.

TERMINOLOGY AND STUDY DESIGN FACTORS

Before we get into the "nuts and bolts" of test methods and examining their results, we must define some basic terms and review a few fundamental concepts in toxicology, combustion science, and the interactions and details which have caused particular approaches to be taken.

Basic Toxicology

The entire purpose of the many combustion toxicology studies performed to date is to evaluate the hazards to man of plastics in real fires. But very few large scale, real fire tests have been performed, and for both ethical and moral reasons none of the studies to date have been done in man. Rather, we use rats or mice as models for man – an approach with some major limitations. The physiology of these animals is not always the same as man, so that they do not give us absolutely the same results as would be seen in man. And, while there are a number of easy ways to determine if a man is incapacitated (that is, if his performance is so impaired that he could not escape a fire), in rodents all we can practically monitor are effects on muscular response and/or coordination.

In monitoring these responses in rodents, the concept of the dose response curve must be kept in mind. This concept is simple - the proportion of animals which display a particular response (death or incapacitation) increases as the "dose" (the amount of decomposition products they are exposed to) increases. That is, higher concentrations will kill more animals (or men) than lower ones will kill. The relationship between concentration (dose) and mortality or incapacitation (response) can be plotted on a curve. In combustion toxicology these curves tend to be very steep, that is, there is usually only a small difference between the concentration that has no effect and that which kills all the animals exposed. From such a curve (or set of curves, one for lethality and one for incapacitation normally being determined for combustion products) the concentration which would be projected to kill or incapacitate 50 per cent of the exposed animals can be determined. These concentrations are called the LC_{50} and IC_{50}, respectively.

As in all inhalation toxicity determinations, the effect of a concentration of respirable material is dependent on the time for which animals are exposed to that concentration. The longer an animal is exposed, the lower a concentration required to achieve an effect. Though the generalization is only good under limited conditions, this relationship (Haber's rule) is expressed as $CT = K$, where C is concentration, T is time and K is a constant particular to the material under study.

Biological systems are affected by a number of factors, and the interactions of these factors can be quite complicated. Different components which can cause death or incapacitation, either singly or in combination, are present in a fire. These are toxic gases and aerosols, oxygen depletion, and heat stress. As the problem of evaluating the interactions between these three components would be very complex (probably impossibly so in a practical sense), the scientific community has concurred with one of the recommendations made by the National Research Council (1977, 1978). Studies to date have been designed as single effect assays - the temperatures that animals have been exposed to and the available oxygen levels (to the animals) have been artificially controlled so that only the actions on the levels of toxicants on animals are studied. There is some justification in holding animal temperature and oxygen levels relatively constant, as these two variables are common to all fires, no matter what the fuel involved.

Different species (such as rats, mice, and man) and, in fact, different strains of the same species (such as Fischer 344 and Sprague Dawley rats) frequently have quantitatively different (and sometimes qualitatively different) reactions to toxicants. These are due to differences in physiology, anatomic structure and the absorption and metabolism of toxicants. A physiological difference between man and the two rodent species commonly used in combustion toxicology is that rodents normally are nose breathers, and possess a fairly efficient particles filtering system in their noses.

Most researchers use rats as their experimental models, but some work has been done with mice, and at least two strains of rats are commonly used - the Fischer 344 and the larger Sprague-Dawley. Each of these differences (species and strain) will lead to some variability in results, and should be kept in mind.

Combustion Terminology

Descriptions of the development and progress of a fire (and the factors which control these) are readily available (Woollsey *et al.*,) but a somewhat specialised terminology has developed in combustion toxicology to describe the different forms of thermal decomposition and some of the related characteristics of materials.

Smouldering is defined as the sustained low temperature combustion of a material in the presence of limited oxygen. Such temperatures are still above the auto ignition point.

Pyrolysis, as defined by Alarie (Alarie and Barrow, 1977), is the thermal decomposition of a material at a temperature below the auto ignition temperature.

Flaming, is the initially highly efficient burning of a material that occurs above the auto ignition temperature in the presence of sufficient oxygen.

Thermolysis is a generic term which describes all the above different forms of thermal decomposition of a material, and has the advantage that, unlike the other terms it does not describe a "pure" condition which rarely occurs alone.

The auto ignition temperature is a specific characteristic of each material. It is generally defined as the temperature at which a material will sustain a flaming condition.

Initially, (until approximately 1976 or so) it was widely believed that smouldering conditions would produce products of greater toxicity that those produced under flaming conditions. This (along with a number of other generalizations) has not proven to be valid. As a single example, take the results of Hilado and Casey, in which they found that as char yield in plastics increased, flamability decreased. For the case of sulfur- and nitrogen-containing plastics, this was found to increase the toxicity of the thermolysis products. But for most of the other plastics evaluated, decreased flamability actually decreased toxicity.

STUDY SYSTEM COMPONENTS

Each of the approaches utilized to date to evaluate the combustion toxicity of materials is based upon a system which can be divided into four components. These are: 1) a device for combustion of the material; 2) a chamber for exposure of test animals; 3) toxicological measurements of animal responses; and 4) chemical analyses of the chamber atmosphere for selected gases and of the blood of test animals for COHb. There are marked differences among test methods in the combustion device, the mode of exposure of test animals and the measurements used for assessment of toxicitybetween each of the principal test methods.

Combustion Devices

The most commonly used methods for thermally decomposing materials in laboratory toxicity studies utilize the tube furnace, the crucible ("cup") furnace or the radiant heat device.

The Tube Furnace

In methods which utilize the tube furnace, the sample of material in a glass tube or other container is thermally decomposed in a furnace equipped with a temperature controller and the products of combustion are carried to an animal exposure chamber by means of a stream of air passing through the tube. The combustion atmosphere may be cooled and diluted by mixing with an additional air stream prior to entry into the exposure system. The exposure chamber, in conjunction with the tube furnace, may be operated in either the static mode (FAA and NBS methods), the dynamic mode (DIN, CMIR, and University of Pittsburgh methods) or both modes (USF). Variations in this basic scheme also have been used in combustion toxicity studies. For example, in the DIN 53 436 method a movable annular electric oven tightly encloses a section of a quartz combustion tube and moves along the tube and test material at a fixed rate, thermally decomposing a constant quantity of the material per unit time (Kimmerle, 1974). Cited as advantages of the tube furnace system by researchers are accurate temperature control, the capability to monitor weight loss, versatility in thermally decomposing materials at a fixed temperature or by ramping the temperature at a fixed rate (Clarke *et al.*, 1982; Kaplan *et al.*, 1983). Disadvantages of the tube furnace are the limitation of the system to relatively small samples and the potential loss of smoke components on the walls of the combustion tube.

The Crucible (Cup) Furnace

The crucible furnace, also referred to as the cup furnace, basically consists of a quartz beaker, crucible or other container mounted in a furnace equipped with a temperature controller. In the NBS test method, the cup furnace is interfaced with the bottom of the exposure chamber, which allows the smoke to be transported into the chamber by convection (Levin *et al.*, 1982). Advantages of the cup furnace are stated to be that larger sample sizes can be thermally decomposed in the cup than in the tube furnace, that there is less opportunity for loss of smoke components than with the tube furnace and that the cup furnace provides close control of decomposition temperature (Clarke *et al.*, 1982). Disadvantages are that air flow into the cup appears to be limited, heat transfer to low density samples is inefficient and there is no provision for continuous monitoring of weight loss of the sample.

Radiant Heat Devices

Radiant heat devices have been of several different designs, but all of these devices thermally decompose materials by radiating infrared energy to the test materials. In one proposed test method, a radiant heat device is interfaced to the bottom of the NBS chamber in place of the cup furnace (Kaplan *et al.*, 1983). Although this method has been used to evaluate the toxicity of combustion products, the test results have not been published in the literature. Cited as advantages of radiant heat devices are that they provide more realistic combustion of composite or multi-layer materials than other devices and that they allow materials to be tested in their end-use configuration and the weight loss of materials to be continuously monitored. Disadvantages of radiant heat devices are that the same heat flux may produce different surface temperatures with different materials because of variations in color and surface temperature, smoke may interfere with transfer of radiant heat energy and there may be a severe shortage of oxygen during combustion.

There has been considerable controversy over which combustion device is the most appropriate for use in evaluating the toxicity of the thermal decomposition products of materials. Investigators have often suggested that the relevance of the combustion conditions to real fires is the most appropriate selection criterion. Others are of the opinion that the utility of the combustion method depends largely on what the test method is intended to accomplish (Clarke *et al.*, 1982). Although none of the combustion devices can simulate all stages of a real fire or all fire scenarios, proponents of each type can identify similarities of the method to some phases of a fire or to certain fire scenarios. Those advocating use of radiant heat devices point out that the primary mode of heat transfer in a real fire is radiation of infrared energy. There is some recent experimental evidence that the cup furnace and the tube furnace can yield combustion atmospheres similar to those present in certain stages of full-scale laboratory fires. In these experiments, the investigators reported that the relative concentrations of selected smoke components produced by the tube furnace were comparable to those in the later, well-developed stages of the full-scale fire (Edgerly *et al.*, 1982). The chemical profile generated by the cup furnace compared well with that in earlier stages. These experiments were limited to the combustion of wood, polypropylene and polymethylmethacrylate.

In test methods using the tube furnace for combustion, the furnace can be operated at one or more fixed temperatures or can be ramped through a range of temperatures at a fixed rate. In contrast, the cup furnace is operated only at fixed temperatures. Investigators who advocate temperature ramping to decompose materials emphasize that this mode of heating simulates the development of many real fires. Decomposition of materials at one or more fixed temperatures, in the opinion of others, allows the evaluation of materials under "worst case" combustion conditions or at selected stages of a real fire. In the opinion of Alarie (1985), the fixed temperature and ramped temperature systems of combustion are equally valid for evaluating the potential toxicity of materials.

Exposure Chambers

The parameters of the exposure chambers used in the various test methods are determined by the mode of exposure (dynamic versus static, whole body versus head only), the rodent species, the type and location of the combustion device and the toxicity measurements. There is considerable variation in the size, configuration and construction materials of these chambers. For example, the exposure chamber of the University of Pittsburgh test method is a 2.3-L glass cylinder, whereas the NBS method uses a 200-L polymethylmethacrylate (or polycarbonate) rectangular chamber. With most methods, test animals are exposed in the head-only mode, to reduce heat stress and to prevent ingestion of combustion products deposited on the fur of the animals. Additional advantages of this mode are that it facilitates blood sampling of animals and respiratory monitoring during the exposure.

The principal difference in the design and operation of the various chambers is determined by whether the test animals are exposed under static or dynamic conditions. In the static mode of exposure, the combustion products enter the chamber and remain until the exposure is terminated. In the dynamic mode, there is a continuous flow of combustion products from the furnace into and through the chamber. Proponents of each mode can identify similarities of the mode to certain real fire scenarios or to some stages of a real fire. It has been suggested that the static exposure (NBS, USF methods) simulates fire conditions in which there is limited flow-through ventilation so that combustion products accumulate in or near the room of origin (Clarke *et al.*, 1982). In contrast, the dynamic exposure systems (DIN, CMIR and University of Pittsburgh methods) simulate fire environments in which victims are exposed to a continuous flow of combustion products. In the DIN fixed-temperature method, the composition of the atmosphere remains constant, whereas, under the ramped heating condition of the University of Pittsburgh method, the composition of smoke to which animals are exposed changes during the exposure.

Toxicity Measurements

The toxicity measurements of the various test methods are almost as diverse as the combustion devices and the exposure chambers. The measurements have been made by techniques that range from simple observations of test animals to monitoring of complex physiological and behavioral responses. With most test methods, incapacitation and lethality of either mice or rats have been the most commonly used indices of toxicity. The University of Pittsburgh test method also provides for the evaluation of sensory irritation by measuring the respiratory rate depression of mice (Barrow *et al.*, 1978). In addition, the electrocardiogram of the rat has been monitored during exposure to pure fire gases and smoke, but this physiological measurement has been used to a very limited extent (Gaume, 1980).

Lethality and LC_{50} Values

The primary index of toxicity used by almost all combustion toxicity test methods has been lethality. Whereas lethality is recognized by toxicologists as a customary starting point in toxicity evaluations of chemicals, many combustion toxicologists unfortunately consider LC_{50} value synonymous with toxicity. With some test methods, lethality has been reported as the ratio of the number of dead animals to the number of exposed animals, but generally a concentration response curve and an LC_{50} value have been obtained. The LC_{50}, in classical inhalation toxicology, is defined as the concentration of gas or vapor that results in the death of 50 percent of the test animals. Because smoke consists of a mixture of particulates and gases, the percentage of animals responding should be related to the concentration of the smoke mixture rather than to the concentration of any single gas in deriving the LC_{50} value. This concentration can be calculated by relating the weight loss of the material to the volume of the exposure chamber in a static system or to the total airflow through a dynamic system. In practice, however, investigators using the cup furnace in the NBS method (static mode), generally report the LC_{50} value as the ratio of the quantity of material that kills 50 percent of the animals (the sample charge) to the chamber volume. With the tube furnace used in the dynamic mode, the University of Pittsburgh method reports the LC_{50} value as the quantity of material that kills 50 percent of the animals whereas, in the DIN method, the weight loss of the material and air dilutions are used in deriving and LC_{50} value. Because of these differences among test methods, it is not often possible to compare LC_{50} values obtained with different methods unless the reported data are expressed on a common basis. Just as with all of its other uses, the LC_{50} is no longer appropriate in combustion toxicology. Rather, some other (less mathematically rigorous) estimate of potential lethality should be sought.

Time to lethality also is considered an important toxicity index of combustion atmospheres because survival and escape from a fire environment are critically dependent on the rapidity of the lethal effects of the fire environment (Clarke *et al.*, 1982). A number of investigators have conducted combustion toxicity studies in which animals were exposed to combustion atmospheres until lethality (or some other effect) occurred in all animals. The average times to lethality caused by smoke generated by different materials have been used to compare and rank materials (Crane *et al.*, 1977; Hilado *et al.*, 1976). This approach has been criticized by Alarie (1985) who considers the use of the average time to death of all exposed animals invalid as a measure of toxicity. However, Alarie and his co-workers (Alarie and Anderson, 1981) have used time to death in conjunction with LC_{50} values to compare the toxicity of combustion atmospheres generated by different materials. These investigators proposed a system for classifying the toxicity of materials (when combusted) in comparison to Douglas fir on the basis of their LC_{50} and LT_{50} value. The LT_{50} value is defined as the time at which 50 percent of the animals die at the LC_{50} sample weight, or the weight nearest above it. The LC_{50} and LT_{50} values of a variety of materials are listed in Table 5-1 and the classification of most of these materials in comparison to Douglas fir is shown in Figure 5-1.

Table 5-1. LC_{50} and LT_{50} values and classification of materials (from Alarie and Anderson, 1981)

Abbreviation	Sample Name - Description	LC_{50} (g)	LT_{50} (min)	LCt_{50} (g.min)	Class[a]
PRC materials[b]					
GM 21	Flexible polyurethane foam	12.9	13	168	B
GM 23	Same as GM 21, with fire retardant	10.4	18	187	B
GM 25	High resilience, flexible polyurethane foam	8.3	19	158	B
GM 27	Same as GM 25, with fire retardant	14.4	15	216	B
GM 29	Rigid polyurethane foam	10.4	28	291	B
GM 31	Same as GM 29, with fire retardant	8.2	23	189	B
GM 35	Rigid polyurethane foam, fluorocarbon blown	7.5	17	128	B
GM 37	Same as GM 35, CO_2 blown	8.0	15	120	B
GM 41	Rigid isocyanurate foam	6.4	18	115	B
GM 43	Same as GM 41, contains some polyurethane	6.1	16	98	B
GM 47	Polystyrene expanded	5.8	11	64	B
GM 49	Same as GM 47, with fire retardant	10.0	9	90	B
GM 57	Phenol formaldehyde - phenol resin, expanded with blowing agent	6.3	20	126	B
Non-PRC materials					
PTFE	Polytetrafluoroethylene resin	0.64	8	5	C
PVC	Polyvinylchloride (92% homopolymer)	7.0	10	70	B
PVC-CN	Polyvinylchloride (92% homopolymer) + 5% zinc ferrocyanide	2.3	7	16	C
PCP-CN	Ploychloroprene (92% homopolymer) + 5% zinc ferrocyanide	2.5	6	15	C
ABS-3	Standard acrylonitrile/butadiene/styrene	6.3	9	57	B
mod.	Modacrylic	4.9	18	88	B
wool	Wool fibers - undyed	3.0	27	81	B
UF	Urea formaldehyde foam	2.5	22	55	B
Cellulose	Blowing type cellulose fiber insulation	11.9	21	250	B
D. Fir	Douglas fir	63.8	22	1404	A
Fiberglass	Fiberglass building insulation, 3.5 inches thick with paper and vapor barrier	35.7	25	893	A
P.E.I	Polyester resin - commercial acrylic modified unsaturated	34.8	14	487	B
P.E.II	Polyester resin - experimental acrylic modified unsaturated	57.4	18	1033	A
H.P.E.	Polyester resin - styrenated halogen modified	14.4	16	230	B
SPF Wood	Compressed spruce, pine, fir slab	48.7	19	925	A

[a] from Figure 5-11

[b] Obtained from the Product Research Committee (PRC) sample bank at the National Bureau of Standards

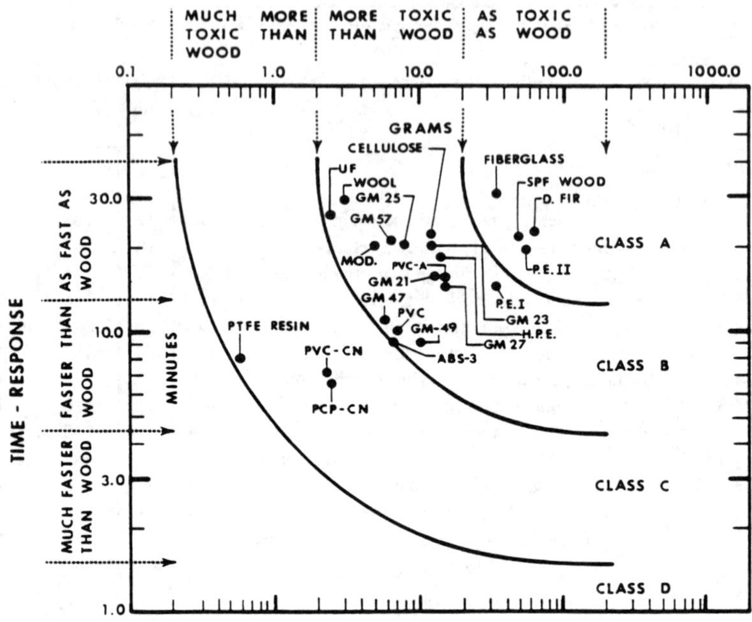

Figure 5.1

Incapacitation

The key to survival in a fire is rapid escape. Delay in, or prevention of, escape leads to further inhalation of toxic smoke and continued exposure to other hazards, and possibly subsequent injury or death. The potential impairment by smoke of mental and motor functions that are necessary for escape was recognized by the National Academy of Sciences in its recommendation that combustion toxicity test methods should include measurements of the loss of capacity to escape and of other behavioral effects (6). A variety of behavioral and physiological methods were developed and/or applied to combustion toxicity tests during the 1970's in order to obtain a measure of the potential of smoke generated by a material to incapacitate an animal. These methods have been reviewed in detail by Kaplan and Hartzell (1984) and by Clarke *et al.* (1982).

Behavioral methods that have been used to measure the incapacitating effects of gases and/or smoke include leg-flexion shock avoidance response (Packham *et al.*, 1978), the motor-driven activity wheel (Crane *et al.*, 1977), the pole-climb conditioned avoidance/escape response (Dilley *et al.*, 1979), the rotarod (Hartung *et al.*, 1977) and the shuttlebox (Boettner *et al.*, 1978; Kaplan *et al.* , 1985). In addition, incapacitation of animals has been evaluated by physiological and observational techniques. Of the behavioral methods, the leg-flexion shock avoidance response has been used most widely and has the largest data base. It is the only method that allows measurement of incapacitation in animals exposed in the head-only mode and location of the behavioral testing apparatus outside the exposure chamber. For this method, the rat is positioned in a tubular restrainer such that an electric shock (approximately 4.5 mA) is delivered to one hind leg upon contact with a metal platform suspended below the leg. Approximately 10-15 minutes of training are required for the rat to learn to avoid the shock by raising its leg above the metal platform. When the animal is unable to avoid the shock during an exposure, incapacitation is deemed to have occurred. As many as six rats have been instrumented to measure incapacitation in smoke toxicity tests.

The second most widely used method is the motor-driven exercise wheel (also called the activity wheel or tumble cage). Although the size and configuration of the equipment vary, the apparatus basically consists of a wire-mesh cage in the form of a wheel which is rotated at a slow constant speed by an electric motor. The animal instinctively moves in a direction counter to that of wheel rotation in an attempt to remain upright in the cage. When the animal is no longer capable of maintaining its movement and begins to slide or tumble as the cage is rotated, incapacitation is considered to have occurred.

Behavioral methods have been used to study the incapacitating effects of pure gases, including CO, HCN, HCl and acrolein, as well as the effects of combustion atmospheres produced by materials. With a few of these methods, concentration-response curves, IC_{50} values (the concentration that incapacitates 50% of the exposed animals) and concentration-time curves have been derived for incapacitation by some of the major toxicants. These data are useful in establishing whether a particular toxicant in a combustion atmosphere is responsible for the incapacitating effects of the atmosphere. In many studies, however, behavioral methods have been used to obtain time-to-incapacitation data for the smoke generated by a fixed quantity of a material and to compare and rank materials by incapacitation times. It is doubtful whether these data, without additional toxicity information, serve any meaningful purpose.

In order to evaluate the sensitivities of the various behavioral methods for measuring the incapacitating effects of CO and HCN, Kaplan and Hartzell (1984) utilized literature data for incapacitation by these gases in rodents and nonhuman primates. Because of the variability in the form of reported data,

Table 5-2. Exposure intensity (C•t Product) of CO associated with incapacitation (Adapted from Kaplan and Hartzell, 1984)

Subject/Method	CO C•t[a] (ppm-min)
Baboons (SwRI)	35,000 (active)
Monkeys (Huntingdon)	25,000 (active)
Rats[b]	50,000 (restrained)
Leg flexion (Utah)	30,000-40,000
(SwRI)	30,000-40,000
Activity Wheel (FAA)	37,000
(Michigan)	30,750
(McDonnell Douglas)	22,000-36,000
Pole-climb avoidance (SRI)	48,000
Rotarod (Michigan)	31,500
(SwRI)	36,000-44,000
Shuttlebox (Michigan)	41,000-53,000
(SwRI)	33,900

[a] C•t products are presented as single values, rather than ranges, when data for only one concentration or time were available or an EC_{50} were reported.

[b] Animals are active in all methods except leg-flexion which requires animals to be restrained

concentration-time (C•t) products were calculated from reported concentrations and exposure times necessary for incapacitation in order to establish a common basis for comparison. The data (Table 5-2) indicate that the sensitivities of most of the methods used to assess incapacitation of rats by CO are comparable and that incapacitation generally occurs after an exposure intensity or magnitude of approximately 30,000 to 40,000 ppm-minutes. This exposure intensity for incapacitation is approximately one-third of the CO C•t product necessary to produce lethality in rats (Kaplan and Hartzell 1984). These investigators also reported some C•t products associated with incapacitation by HCN but the data were too limited and variable to compare the sensitivities of the behavioral methods.

Although behavioral methods are sensitive to the incapacitating effects of hypoxia-producing agents, this methodology appears to have limited utility in evaluating the incapacitating effects of irritant gases. Data obtained from experiments in which rats were exposed to either pure HCl or acrolein show that rats are extremely tolerant to irritant gases. When concentrations of these gases are sufficiently high to incapacitate animals, the animals die within a short time afterward (Kaplan *et al.*, 1985; Crane *et al.*, 1985). Even those concentrations which are not high enough to incapacitate rats may cause a high incidence of mortalities during the usual 14-day post-exposure period. This tolerance of the rodent to irritant gases may be a serious limitation to its use in combustion toxicity tests.

Physiological indices that have been used to assess the incapacitating effects of gases and/or combustion atmospheres include the depression of respiratory rate and the onset of cardiac arrhythmia. Most studies of the effects of gases and smoke on the inhibition of respiration have been conducted at the University of Pittsburgh and are reviewed in the next section on sensory irritation. In the only study of the utility of cardiac arrhythmias in assessing incapacitation, Gaume *et al.* (1981) reported that arrhythmia occurred just prior to incapacitation of rats by CO in an activity wheel. In the University of San Francisco test method, incapacitation has been assessed by observation of the loss of equilibrium (staggering), prostration, convulsions or collapse (Hilado *et al.*, 1976). However, observational techniques are highly subjective and, in general, are less desirable than other methods.

Sensory Irritation

Sensory irritants are prevalent in combustion atmospheres and may play a significant role in fire fatalities attributed to smoke inhalation. These chemicals irritate the eyes, causing pain, blinking and lachrymation, which may interfere with vision. They also affect the upper respiratory tract, causing a reflex inhibition of respiratory rate and a burning sensation in the nose, mouth and throat. Severe irritation may result in inflamation, swelling of the larynx and trachea, laryngeal spasms and bronchoconstriction (Alaine, 1966). It is quite possible that the effects of sensory irritants delay the escape of individuals from the fire environment and even incapacitate them so that continued inhalation of CO would result in high blood COHb levels and death would be attributed by postmortem examination to CO.

The only test method that provides a direct measurement of the sensory irritation of combustion atmospheres is the University of Pittsburgh method (Barrow *et al.*, 1976). The measurement is based on the reflex inhibition of respiratory rate caused by stimulation of trigeminal nerve endings in the nasal mucosa. Respiratory rate is monitored in four mice in body plethysmographs during exposure and a concentration-response relationship for respiratory depression and an RD_{50} value are determined by varying the quantity of material decomposed in successive experiments. The RD_{50} is defined as the quantity of material that produces a 50-percent decrease in respiratory rate. It has been claimed that the RD_{50} concentration in mice is equivalent to an intolerable level of sensory irritation for humans and will incapacitate a human within 3 to 5 minutes (Alarie, 1973; Kane *et al.*, 1979). Data obtained in human volunteers exposed to the smoke from Red Oak are not consistent with this claim (Potts and Lederer, 1978), and hence it is possible that some factor of the RD_{50} value may be useful in predicting the incapacitating concentration of gases and smoke in humans. For example, Alarie (1981) correlated RD_{50} values with ACGIH TLV-TWA values for 25 chemicals and proposed that the 0.03 RD_{50} value may be useful in establishing TLV-TWA values for airborne chemicals (it should be kept in mind however that, though many of the older TLV's were set based on respiratory irritation, this is no longer the case).

Other Toxicity Indices

In later studies by Alarie and his co-workers (Alarie and Anderson, 1981; Barrow *et al.*, 1979), the University of Pittsburgh test method was modified and expanded to provide a more comprehensive evaluation of the toxicity of combustion atmospheres. In addition to sensory irritation, measurements of stress, lethality, asphyxiation and histopathological changes were included in the test protocol. According to Alarie and his co-workers (Alarie and Anderson, 1981) sensory irritation evokes a series of physiological adjustments, including changes in heart rate, blood pressure and blood vessel size, to compensate for the reduced breathing rate and apneic periods. Therefore, a stress index (SI), calculated from the rate of onset of, and recovery from, the respiratory effect and the degree of respiratory depression, was used as an index of the stress of a combustion atmosphere. A concentration-response relationship was established for physiological stress and a SI 100 value (the sample size associated with a stress value of 100) was derived. The SI 100, RD_{50} and LC_{50} values of combustion atmospheres produced by various materials (Table 5-3) were used to classify materials in comparison to Douglas fir. Asphyxiant effects were evaluated from the characteristic alterations of the pattern of the animal as well as by the absence of body movements. An asphyxiant range, *i.e.*, the sample weight range in which asphyxiation was observed, was reported as an index of toxicity. Finally, histopathological examinations were made of the major organs of exposed animals sacrificed at 24 hours following exposure. Numerical ratings were used to indicate the severity of damage to the upper respiratory tract and, in other organs, the absence or presence of lesions was noted.

Analytical Methods

Although biological approaches have been predominant, a few investigators have proposed chemical or analytical approaches to evaluate the toxicity of combustion atmospheres. Sumi and Tsuchiya (1975), for example, defined a "toxicity" factor, $t = C/C_f$, for each gaseous component of smoke, where C is the concentration of the gas in smoke and Cf is the concentration that is fatal to man in 30 minutes. The total toxicity of a combustion atmos-

Table 5-3. Classification of thermal decomposition products from tested materials[a] in comparison to wood [b] (Adapted from Alarie and Anderson, 1979)

Toxicity indices measured as	Classification			
	Much more toxic than wood	More toxic than wood	Similar to wood	Less toxic than wood
Acute Mortality (LC_{50})[c]	LC_{50} = 0.2-2.0g PTFE	LC_{50} = 2-20 g All materials except PTFE and wood	LC_{50} = 20-200 g None found[d]	LC_{50} > 200 g None found
Physiological Stress (SI100)	SI100 = 0.012-0.12 g None found	SI100 = 0.12-1.2 g GM 29, GM 31, GM 35, GM 37, GM 41, GM 43, GM 57, UF, PVC-A	SI100 = 1.2-12g GM 21, GM 23, GM 25, GM 27, GM 47, wood	SI100 12g GM 49
Sensory Irritation (RD_{50})	RD_{50} = 0.1-1.0 mg None found	RD_{50} = 1.0-10 mg GM 21, GM 23, GM 25, GM 27	RD_{50} = 10-100 mg GM 31, GM 35, GM 39, UF, PTFE, PVC-A	RD_{50} > 100 mg GM 29, GM 41

[a] Materials are identified in Table 5-1
[b] Wood taken as Douglas fir
[c] Calculated on the basis of deaths recorded during the 30 minute exposure and 10 minutes of recovery
[d] Authors' note: Fiberglass (LC_{50} = 35.7) should be included in this category

phere is obtained by summation of the t factors for the gases in the atmosphere. Tsuchiya (1981) subsequently expanded this concept in developing a "dynamic toxicity factor," TD, which takes into account the rate of production of toxic gases per unit surface area of the material as well as the toxicities of the gases. An analytical approach was also developed by Spurgeon (1978) who correlated the times to incapacitation of rats with chemical analyses of the combustion atmospheres of 75 aircraft interior materials. A mathematical expression was developed for calculation of incapacitation time based on the yield of nine combustion gases, CO, HCN, H_2S, NO_2, SO_2, HCl, CH_2O, HBr and HF.

All methods based solely on chemical analyses have the same major deficiency, which is that the prediction of potential toxicity from chemical analyses of common toxicants is not always reliable. Two examples of materials which produce combustion atmospheres whose toxic effects are not predictable from analyses of the atmospheres are Teflon 100® (fluorinated ethylene/fluorinated propylene) and Tefzel 200® (ethylene-tetrafluorethylene). In a study of the combustion toxicity of halogenated polymers, Kaplan and others (1984) reported that insufficient quantities of CO, HF and COF_2 were produced by these two materials to account for the observed toxicity. Even when the major toxicants in a combustion atmosphere can be identified, the prediction of toxicity from these analyses may be speculative because of possible biological interactions between these chemicals. Therefore, the use of animals is necessary in combustion toxicity tests of materials to detect the presence not only of unusual or uncommon toxicants but also of biological interactions between common gases.

METHODS FOR EVALUATING THE BIOLOGICAL HAZARDS OF COMBUSTION PRODUCTS

Objective of, and Requirements for, Evaluation Methods

The primary objective of any method for evaluating the biological or toxicological hazards of exposure to a chemical is the determination of the chemical's potential to cause adverse effects in humans. In the case of chemicals present in the industrial and occupational environments, data from epidemiological studies and industrial and occupational human exposures often are sufficient for this assessment. When such data are not available, toxicological studies in which animals are exposed to the chemical of concern must be conducted. These studies must identify the nature of the toxic effects, the target organs affected and the mechanism of action of the chemical to enable a valid extrapolation to humans. This same requirement is applicable to methods for evaluating the biological hazards of combustion atmospheres produced by thermal decomposition of materials.

Screening *Versus* Evaluating

In recent years, there has been considerable controversy in the fire science community over the need for a standardized "toxicity screening test." In addition, proponents of a screening test have different views regarding what the test should accomplish. Some are of the opinion that the purpose of the test should be to identify those materials that generate "highly" or "unusually" toxic combustion atmospheres. Although the criterion for unusual toxicity is not clearly defined, there is universal agreement that smoke produced by polytetrafluoroethylene (PTFE) is representative of that category since LC_{50} values for all other materials tested to date are at least an order of magnitude greater than that of PTFE smoke. Proponents of the "unusually toxic" screening test generally claim that these other materials should "pass" the test because none of their LC_{50} values differ by more than a factor of ten. In view of the fact that a single experiment with any of the available test methods is capable of differentiating between the toxic potency of PTFE smoke and that of other materials, the adoption of this type of toxicity screening test would have minimum economic impact on manufacturers and consumers. The test, however, would not reduce the incidence of fire fatalities or serve any other useful purpose.

Other advocates of a screening test insist that a screening test must yield numbers that can be assigned to materials on the basis of the "toxicity" of their combustion atmospheres. These individuals argue that, because numbers are used to rate certain flamability properties of materials, a comparable numbering system can be established for the toxicity of combustion products for use by code officials and regulatory agencies. The LC_{50} values required under Article 15 of New York State's Uniform Fire Prevention and Building Code (N.Y., 1985) is an example of a numbering system, although the purpose is to make the data available to the public rather than for regulatory action. The assignment of numbers to materials is strongly opposed by plastics manufactures, who contend, and rightfully so, that numbers imply differences in toxicity even though these differences often are not of any significance. In addition, these numbers may be misinterpreted by individuals. To avoid this possible misinterpretation, some advocates of a screening test propose that materials be separated into classes or categories on the basis of the LC_{50} values of their combustion atmospheres. With this system, minor differences in LC_{50} values would not be apparent.

Regardless of whether LC_{50} values or classes of toxic potency are used, any "toxicity screening test" based solely on the lethality of rodents has limited relevance and utility. This type of test only provides a measure of the short-term toxic potency of the combustion atmosphere of a material in rodents. It does not necessarily provide a measure of the toxic potency to humans nor does it evaluate the potential of the atmosphere to cause adverse effects in humans. One of the major deficiencies of a screening test based on lethality is its failure to take into account the nature and rapidity of the toxic effects of a combustion atmosphere and the mechanism of action of the toxicants. For example, it is evident that the toxicity and potential hazard of the combustion atmospheres of the materials discussed in Chapter 7 differ considerably, although their LC_{50} values are generally similar. The toxic hazard of a combustion atmosphere containing HCN and CO is certainly not the same as one containing primarily CO, even though the two atmospheres have comparable LC_{50} values. Similarly, the toxicity and hazard to humans of an atmosphere that is highly irritating and causes post-exposure deaths in rodents are not the same as one in which CO is the major toxicant, despite similar lethal potencies in rodents. Another major deficiency in the use of a screening test based on toxic potency is that the lethality in rodents does not appear to be a suitable toxicity index for evaluating the effects of irritant gases in man. Recent studies of the effects of HCl in primates have shown that the respiratory response of rodents to sensory irritants is markedly different than that of primates and could lead to an inaccurate evaluation of the potential of these gases to cause injury and death in humans (Kaplan *et al.*, 1986).

Approaches to Evaluating the Toxicity of Combustion Atmospheres

The recommended approach for evaluating the potential of combustion atmospheres to cause adverse effects in humans consists of two separate phases. In the first phase, the toxicity of an atmosphere is characterize by animal experimentation and correlated with analytical data in an attempt to identify the major toxicants. When the identities of the major toxicants are established, available human data for these gases can be utilized in the second phase to evaluate the toxicological hazard of the combustion atmosphere. Considerable experience has shown that, in the case of many combustion atmospheres, the integration of toxicological and analytical data can establish the identity of the causative species, particularly when CO and/or HCN are the major toxicants. In addition, sufficient animal data have been obtained in studies with HCl and HF to establish whether these gases were responsible for or contributed to the observed toxicity. In the event the toxic effects of a combustion atmosphere are not consistent with the analytical findings, further analytical work will be required to identify the causative species.

The major limitation to this two-phased approach is that considerable analytical effort may be required to establish the identity of the major toxicants of some combustion atmospheres. In most cases, these toxicants will be irritant gases, perhaps various aldehydes, organic acids, NO_2, nitriles, or isocyanates, which cause irritation of the eyes and respiratory tract and post-exposure deaths in rodents. If the identity of these gases can be established, available human data often are sufficient to enable an assessment of the hazards of the combustion atmosphere to humans. In the event major toxicants are not identified, the results of animal experimentation should be used with caution in evaluating the toxic hazard of these atmospheres to humans. Another limitation to this or any approach is our limited knowledge of the combined effects of hypoxia-producing gases, such as CO and HCN, or of irritant and hypoxia-producing gases in humans. Although the biological interaction of some combustion gases have been investigated in rodents, the results of recent studies with nonhuman primates indicate that investigation of the combined effects of combustion gases in primate species is warranted.

Applicable Concepts

Concentration-Time Product

In a classical inhalation toxicity study of a gas or vapor, animals generally are exposed to a constant concentration of the chemical for a specified period of time. The exposure is described in terms of the concentration and the time of exposure. In contrast, in combustion toxicity experiments, animals are exposed to a mixture of gases, the concentrations of which are not constant during the usual 30-minute exposure period because these gases are evolved at different rates. It is possible to describe the intensity or magnitude of exposure to the major components of the smoke in terms of a concentration time product (C•t), provided that these gases are continuously monitored or that frequent samples are taken and analyzed. The concentration-time product of a gas is equal to the integrated area under the concentration-time curve and may be expressed in units of ppm-minutes, mg/L minutes or mg/m^3 minutes. It also is possible to describe the intensity of exposure to the total combustion atmosphere, rather than to individual gases, in terms of a C•t product calculated from the weight loss of the combusted material with time (Alexeef and Packham, 1984).

As pointed out by MacFarland (1977), the C•t product cannot be the toxicant dose because the dose is an amount or quantity of material. However, it is a measure of the magnitude or intensity of the exposure conditions and does bear some relationship to the dose. As either the concentration of the gas or the time of exposure increases, the quantity of material that enters the respiratory tract increases (assuming ventilation does not decrease) and, therefore, the dose increases. The more general issue of how such short term "ramping" (running up and then down) of toxicant concentrations alters biological responses is an unaddressed one.

Haber's Rule

Haber's rule states that the C•t product necessary to elicit a fixed response is constant, *i.e.*, C•t = K (MacFarland 1977). Thus, according to the rule, if C_1t_1 causes 50-percent mortality, C_2t_2 will also result in death of half of the exposed animals provided $C_2t_2 = C_1t_1$. When tested experimentally, the rule has been found to hold reasonably well for a limited range of C or t. The range over which the C•t product is constant occurs in the curved portion of the hyperbolic concentration-time curve for a response. Deviations from the rule occur with both high and low concentrations of a gas (and short and long exposure times) so that the ends of the hyperbola become asymptotic. Despite these deviations, C•t products are useful in evaluating whether major toxicants, such as CO, HCN and HCl, are responsible for certain observed effects (incapacitation and lethality) of combustion atmospheres. For these evaluations, the C•t products necessary to produce these effects must be determined in pure gas experiments.

Methodology

Recommended Test Method

All of the combustion toxicity test methods have advantages and limitations. The University of Pittsburgh and NBS test methods have considerable data bases and, in the opinion of the author, are the currently accepted methods of choice for evaluating the toxicological hazards of combustion atmospheres. Because considerably more pure gas data and analytical results have been reported with the NBS method, the NBS method, with modifications, is recommended as the test method for the approach designated in this chapter. The NBS test method is described in detail in NBSIR 82-2532, "Further Development of a Test Method for the Assessment of the Acute Inhalation Toxicity of Combustion Products" (Levin, 1982). Recommended modifications to the method include measurements of incapacitation and of sensory irritation, histopathological examination of animals and additional chemical analyses.

Combustion

The conductive type furnace (Figure 5-1) described by Potts and Lederer is utilized as the combustion device. Nonflaming combustion experiments are conducted at 25°C below the predetermined autoignition temperature; flaming combustion experiments are conducted at 25°C above this temperature, except when high temperatures may be necessary to maintain flaming combustion. The autoignition temperature of a test material is established in a series of experimental trials without animals, by decreasing or increasing the furnace temperature from an initial estimated starting temperature, with an upper limit of 800°C.

Exposure Chamber

Test animals are exposed to combustion atmospheres in a nominal 200-L exposure chamber (Figure 5-2) with interior dimensions of 122 × 36 × 46 cm and constructed of clear polymethylmethacrylate or polycarbonate. Animal ports constructed of polymethylmethacrylate tubing permit exposure of six rats in the head-only mode to the smoke generated within the chamber.

Animal Exposures

Adult male Sprague-Dawley rats, weighing between 250 and 350 grams, are used to characterize the toxicity of the combustion atmospheres generated by a material. Prior to the start of each experiment, six animals are weighed and placed in tubular aluminum restrainers which are inserted into the chamber. The experiment is initiated with the addition of a weighed sample of the test material into the tared quartz beaker after equilibration of the selected furnace temperature. In each experiment, six animals are exposed to the combustion atmosphere for a total of 30 minutes. At least five experiments are conducted with varying quantities of the material in both the flaming and nonflaming combustion modes in order to derive concentration-response curves and LC_{50} values. The sample weight used in an experiment is related to the chamber volume and expressed as the sample "charge" or sample concentration in mg/L units.

Toxicity Measurements

Incapacitation

Five of the six animals are used for the measurement of incapacitation by the leg-flexion shock avoidance response. Each animals's responses and time of responses are recorded on a strip chart.

Lethality

The five test animals in which incapacitation is measured are observed for lethality during exposure and for 14 days exposure. Animals that expire during exposure of the first hour post-exposure are considered exposure lethalities; those that expire after the first post-exposure hour are considered post-exposure lethalities. The concentration-response relationship and slope are established for lethality and an LC_{50} value with confidence limits (30 minute exposure and 14-day post exposure) is derived by the probit method Finney (1971) or by another appropriate statistical method.

Blood COHb Analyses

The sixth animal is used to obtain a measure of the percent COHb saturation at the median time of incapacitation. immediately after incapacitation of the third of the five test animals, the sixth animal is removed and a blood sample is obtained by intraorbital venous puncture. In the event none of the animals are incapacitated, a blood sample is obtained from the sixth animal at the end of the exposure. Blood samples also are obtained by cardiac puncture from all animals that expire during exposure.

Toxic Signs and Pathology

Animals that survive the exposure are observed for toxic signs such as respiratory abnormalities, ataxia and loss of righting reflex during the first hour post-exposure and periodically for 14 days. Animal body weights are recorded pre-exposure and on days 1, 7 and 14 post exposure for comparison with body weight growth patterns of control animals. Animals that die during exposure or post-exposure are necropsied and major organs (respiratory tract, heart, kidneys, liver and spleen) are examined grossly and by light microscopy.

Combustion Atmosphere Analyses

Analyses of the combustion atmosphere during exposure are made continuously for O_2, CO_2 and CO, and either continuously or at frequent intervals for other major toxicants to enable calculation of C•t products for these toxicants. Analyses are performed for HCN, HCl and HF as warranted by the chemical structure of the material. Analyses for additional compounds, including aldehydes, nitriles and isocyanates, may be required to establish the identify of major toxicants in the combustion atmospheres generated by some materials.

Sensor Irritation

Although the measurement of sensor irritation is not specified in the NBS test protocol, sensory irritation is, in the opinion of the author, an important toxicity parameter of combustion atmospheres. Relatively simple modifications to the NBS exposure chamber to allow the monitoring of the respiratory response of rats to gases and combustion atmospheres have been described by Hartzell and others (Hartzell *et al.*, 1983). By conducting a series of experiments with different quantities of a material, a concentration-response relationship for respiratory depression and an RD_{50} value may be derived. These experiments generally require smaller quantities of materials than those for determining incapacitation and lethality and cannot be conducted in conjunction with the experiments in which incapacitation and lethality are measured.

Existing Test Systems

Over the years a number of test systems have been developed and utilized to evaluate the toxicity of polymer thermolysis products. Most of the data reviewed later in this chapter come from eight of these systems, which will be described and discussed in some detail. There are (or have been) a number of other systems which, due to their not having generated as large an available data base, we will not consider in detail here. But it should be noted in passing, at least, that other systems (Hofmann and Oeltel, 1968; Kishitani and Nakamura, 1974; Warity, 1975; Wright and Adams, 1977; Rodkey and Jenkins, 1979) have been developed and utilized.

National Bureau of Standards (NBS now the NIST - National Institute for Standards and Technology) Method

This test method, under development since about 1976 at the U.S. National Bureau of Standards, has been described in detail in a Bureau report entitled "Further Development of a Test Method for the Assessment of the Acute Inhalation Toxicity of Combustion Products" (NBS, 1982). An *ad hoc* working committee, consisting of members from approximately 20 academic, industrial or government organizations engaged in relevant work, was formed in 1977 for the purpose of providing a forum for exchanging technical information to assist the Bureau in the development of the test method (Birky *et al.*, 1980). Eight labs from the *ad hoc* working group also participated in a "round-robin" evaluation of the test method as it evolved, with the examination of 12 materials, both natural and synthetic. The announced objectives of the intralab evaluation were to determine the operability of the procedure and to determine the reproducibility of test results from different laboratories, with many of the participants believing other test systems would also be evaluated and a "best" system selected based on the results. Currently, this test system, with variations, is the most widely used test system in the field (though the State of New York's endorsement of the University of Pittsburgh system has made that a close second).

According to the NBS report, the test method "... provides a means of assessing the acute inhalation toxicity of the combustion products of materials under specified laboratory conditions and is primarily intended for research and preliminary screening purposes. Additional factors must be considered in evaluating the potential toxic hazard posed by a material in a given situation... Therefore, the results of this test method must be combined with other information when making decisions about the suitability of materials for specified uses" (NBS, 1982).

The basic components of the NBS test method consist of the following:

The furnace originally employed in this system is the "Potts pot" (Potts and Lederer, 1977) - a conductive box heater as shown in Figure 5-2. The sample is held in a quartz beaker in the middle of this box heater, the temperature being controlled (in an off/on, all-or-none manner) from the sensed temperature of the beaker. The furnace operates in the temperature range of 300-900°C (± 15°C without sample - there is a somewhat greater variation with the sample in place). The protocol employed calls for heating the furnace to the desired target temperature, then dumping the sample into the furnace and sealing the system. As can be seen from Figure 5-3, the furnace is located in the floor of the exposure chamber. The drawbacks to this decomposition system are substantial and numerous. Control over the decomposition conditions (both in terms of temperature and oxygen availability) is minimal. A strong case can be made that the pot can only perform destructive distillation and pyrolysis of samples and connecting the furnace directly to the animal exposure chamber adds a heat load to the system which forces other design factors in compensation.

Analytical monitoring is by continuous IR spectrophotometric measurement of CO, O_2 and CO_2. The overall concentration achieved is calculated, assuming that the entire sample is decomposed, by dividing the weight of the fuel load by the total volume of the system. This approach to calculating concentrations tends to favor those polymers which contain fillers or otherwise act to leave large residues.

The chamber, as shown in Figure 5-4, is a plexiglass rectangular shaped box, having a total volume of 200 L, and ports through which the head only exposure of animals can be accomplished. The shape of the chamber is such that though adequate mixing is accomplished in a short period (several minutes) for gases, a visible layering of aerosols for periods of several minutes is frequently observed.

Six male Sprague-Dawley rats are used for each 30 minute exposure, with two of these animals being terminated immediately at the end of the exposure (to obtain blood samples to determine COHb, O_2Hb and total Hb). Incapacitation during exposure is evaluated in terms of loss of ability to keep the hind feet off a plate, the touching of which induces an electrical shock. The animals which do not die during exposure are observed for 14 days after exposure or until death (whichever comes first).

The protocol which employs this system calls for determinations of concentration response curves (with death as the response) at each of three sample temperatures (the autoignition temperature, and 25°C above and below it).

NBS Furnace ("Pott's pot")

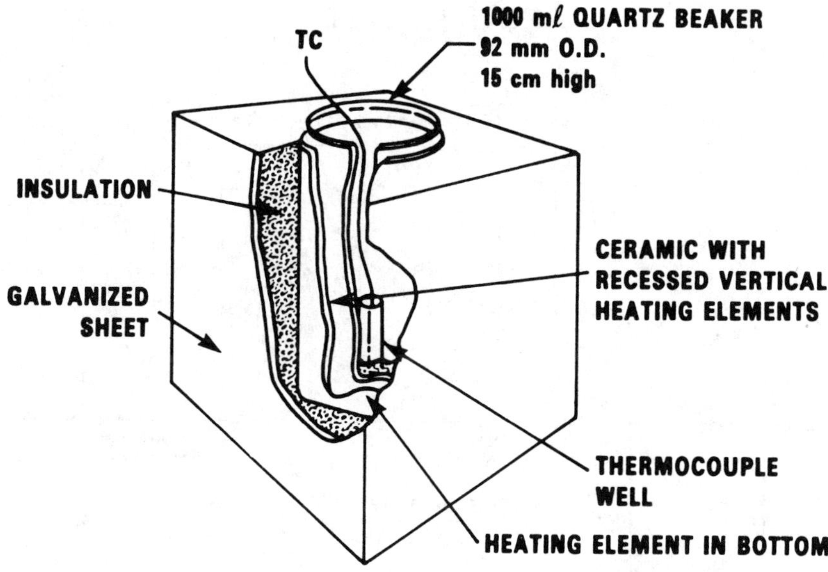

Figure 5.2

A major disadvantage of the original NBS test method relates directly to the combustion device, a type of crucible furnace. Since heat transfer with this furnace is intended to be largely conductive, contiguous contact of a sample with the beaker surface must be maintained. Thus, the furnace is most suitable for thermoplastic materials or samples in a powder or pellet form which will melt and conform to the shape of the beaker. Considering this, it is interesting that the NBS method utilizes a block of char-forming and non-melting material, Douglas fir, as its reference material. Use of a commercial, pelletized thermoplastic material might be more appropriate.

Problems have been encountered with low density foam samples, which filled the beaker and restricted the movement of air around the sample. Furthermore, large samples, such as are needed with low density materials, are subjected to a wide range of temperatures, since the air temperature decreases near the top of the beaker. It is not surprising, therefore, that the

Schematic Of NBS Combustion Toxicity Testing Set Up

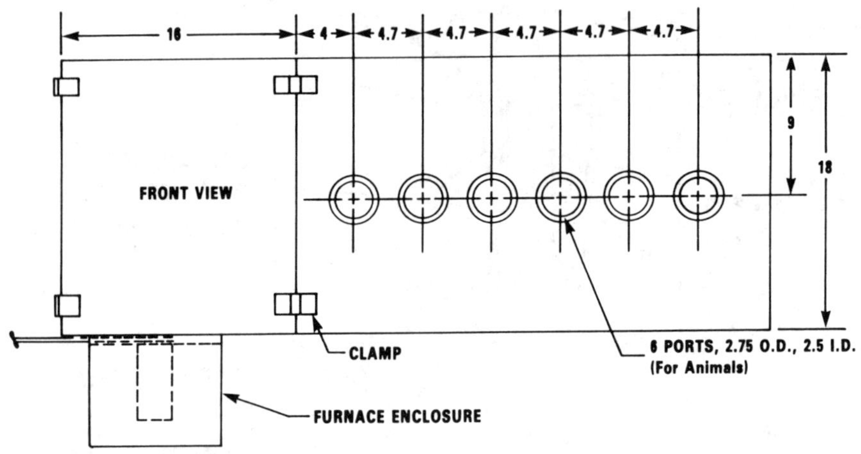

Figure 5.3

reproducibility of results in the case of such foams is not as good as with many other materials. The NBS combustion technique also does not satisfactorily combust, in a consistent manner, other materials such as laminates, coated specimens, carpeting/backing, one-surface fire barriers, foamed thermosets and many other real-world materials.

One response to the limitations of the original NBS method was to utilize a radiant heat energy thermal decomposition device in place of the Pott's furnace, and to include a measure of incapacitation. In this variation, developed by S. C. Packham, the same 200-L exposure chamber and analytical devices are utilized. However, both the combustion furnace and the experimental protocol are considerably different. The combustion furnace (Figure 5-5) was developed by H.W. Stacy in the laboratories of the Weyerhaeuser Company. It has an elongated, trapezoidal shape with quartz windows forming the nonparallel sides. The frame is constructed of stainless steel. The

NBS Exposure Chamber

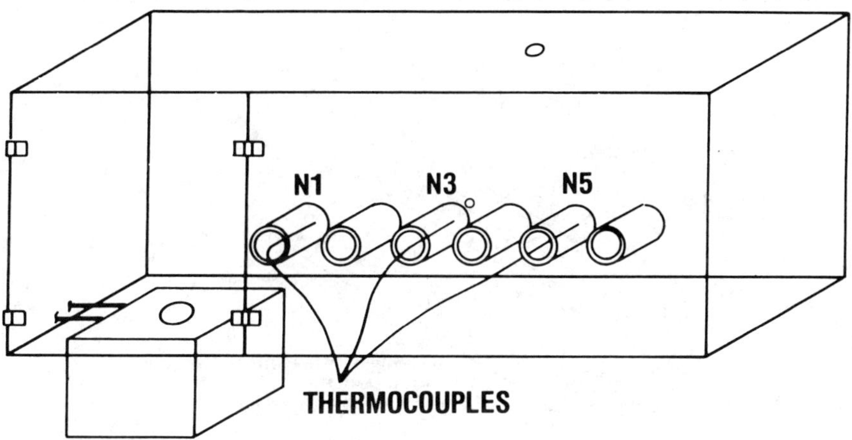

Figure 5.4

radiant heat source consists of tubular quartz, tungsten-filament lamps housed in parabolic reflectors. Using lamps with a 2,000-Watt power rating, heat fluxes up to 5 W/cm^2 can be obtained. A sample holding tray is suspended from a load cell into the combustion zone of the furnace. The furnace configuration allows testing based on exposed surface area for many materials such as composites, laminates, tiles, fabrics and coatings. Procedural dosages are expressed in terms of specimen surface area exposed to a designated heat flux.

Incapacitation is monitored using the leg-flexion shock avoidance method.

Materials are evaluated for both acute toxic potency and general toxic potency on the following bases: acute toxic potency is based on average times to incapacitation (T_is) determined from tests on sample specimens having a surface area of 100 cm^2; general toxic potency reports are based on 14-day

Quartz Windowed Radiant Heat Furnace

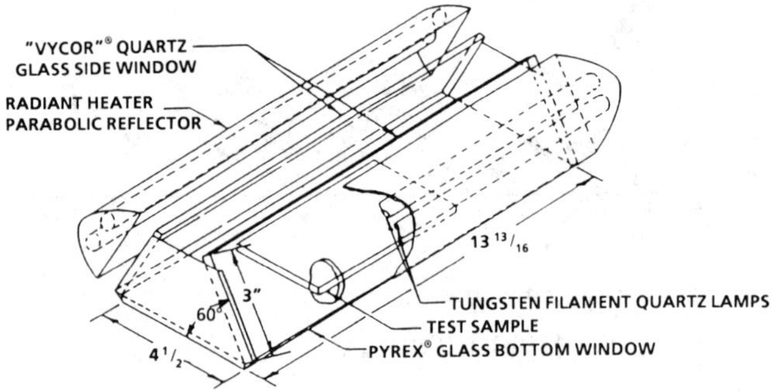

Figure 5.5

lethality and gross lung and respiratory tract pathology determined from tests on sample specimens having a surface area of 20 cm^2. This test method employs radiant heat for combustion of a material, a process which occurs in most fires and fire test procedures. Thus, selection of parameters can be made in order to be of use in fire scenario modeling, *e.g.*, ramped heat flux, fixed heat flux, flaming, nonflaming, *etc.* One advantage of the test method is that it permits the evaluation of many end-use materials (*e.g.*, composites, laminates, coatings, films, *etc.*) which cannot be appropriately accommodated in most of the other test methods. Also, the surfaces of materials are exposed to heat, which is the way materials are likely to be exposed in certain stages of a real fire. The method expresses toxic potency on the basis of surface area (LA_{50}) rather than weight. Another advantage is that the method provides a measurement of time-to-incapacitation.

A disadvantage of this method is that different samples may have variable absorptivities to radiant heat due to surface color, char formation or other characteristics. This may result in different surface temperatures among materials, when the same heat flux is applied. Also, smoke buildup in the combustion chamber may reduce the heat flux at the surface of the sample during the course of the test. Radiant heating may cause a greater temperature buildup in the animal exposure chamber than that produced by other combustion devices. Finally, the selection of a 10-minute incapacitation time as the criterion to differentiate between two different types of test reports is arbitrary and needs further evaluation when sufficient data from the test method become available.

Carnegie Mellon Institute of Research System

This system was developed over the period 1977 to 1979 by J. B. Reid, T. F. Brecht and S. C. Gad with funding from Union Carbide Corporation. It has been described by Gad and Smith (1984).

The furnace utilized is shown in Figure 5-6. A focused infrared heater serves to heat a sample on a quartz plate on a thermogravimetric analysis (TGA) device. A thermocouple directly in the sample serves to monitor sample temperature and can control this for a range from 150°C to 1500°C (± 5°C). Alternatively, the system can control from the heat flux from the IR heater. As shown in Figure 5-7, the furnace is not directly in the animal exposure chamber. Rather, it is connected by a series of glass tubes which serve to close the entire system. A sealed fan acts to circulate and mix the chamber contents, and warming tapes about the glass tubing prevent condensation and collection of aerosols. A Data Trak 5600 microprocessor controls the entire system, allowing complete and precise reproducibility and the ability to alter sample temperature in any desired manner (5°C/min, 10°/min, *etc.*). The furnace is also configured to allow complete control over the initial atmosphere of the furnace (by adding O_2, N_2, or CO_2), and a propane torch acts as an ignition source when necessary. Concentrations are calculated based on the weight of sample volatilized.

Analytical monitoring in this system is based primarily on continuous measurement (and recording) of O_2, CO, CO_2, NO/NO_X (all by IR spectrophotometry), total hydrocarbons (by flame photometry), and HCN (by a specific electrode). Periodic measurements of other, special interest gases can be made by draw tubes.

The animal exposure system is a glass cylinder with stainless steel ends, having a motor driven activity wheel suspended on a stainless steel shaft through the chamber. The volume of the complete system is 55 liters.

CMIR Ignition Chamber System

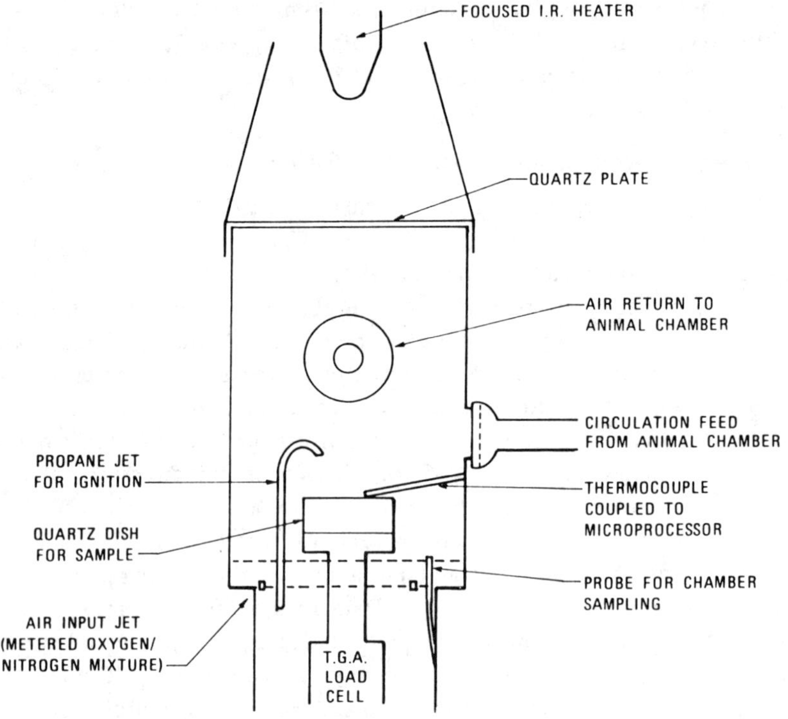

Figure 5.6

Six male Sprague Dawley rats are utilized for each 30 minute exposure. Incapacitation is evaluated as occurring once an animal is incapable of walking on the wheel, but rather tumbles. A high intensity lamp placed behind the chamber provides the ability to observe animals when even a heavy smoke is generated. A disadvantage is that this incapacitation model can be called a subjective one, and indeed it requires some degree of observer training. One

CMIR Combustion Toxicity Testing System

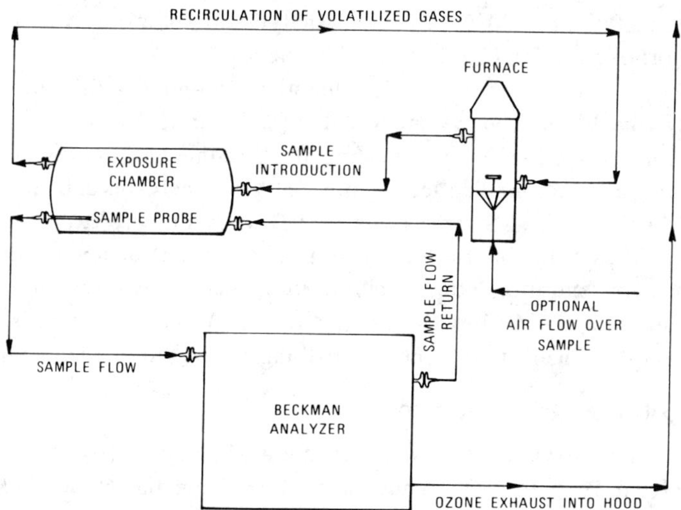

Figure 5.7

approach to adding an objective measure to the system is to monitor the torque or drag on the turning wheel, but this effort has not been pursued to completion. Endpoints are time to incapacitation, lethality and (with blood taken from one animal, by heart puncture after the exposure) COHb, O_2Hb, and total Hb.

The protocol employed calls for determining concentration response curves for both death and incapacitation at each of three temperatures (these temperatures normally being achieved by raising the sample temperature at the rate of 40°C per minute). The three temperatures are 10°C below the AIT, 10°C above the AIT, and 440°C.

The major drawback to this system is the capital equipment cost. It is, however, probably the most versatile and reproducible of the existent systems.

United States Testing System

This U.S. Testing System was developed by Eugene Rider at U.S. Testing during 1976. It utilized as a furnace a conductive heated tube furnace with an on/off controller, controlled from a thermocouple contained in the wall of the quartz furnace tube. The furnace has the capability of heating a sample within the range 200°C to 800°C (± 25°C). Analytical monitoring is by continuous IR monitoring for O_2, CO, CO_2 and NO/NO_X.

The animal chamber is a 52 liter glass cylinder with a motor driven activity wheel in it. The system is sealed (and therefore technically "static"), with a fan serving to circulate the atmospheric contents. No provision is made to avoid condensation or collection of volatiles for aerosols at bends and "cold spots" in the system. Six male Sprague-Dawley rats are exposed for thirty minutes (then held for fourteen days post exposure) at levels sufficient to establish concentration lethality curves under each of two sets of decomposition conditions (25°C above and below the AIT). Incapacitation is evaluated in terms of the inability to continue walking on the motor driven wheel.

University of Pittsburg System

Originally developed by Y. Alarie and C. S. Barrow (Alarie and Barrow, 1977) from 1973 to 1976 (under a contract from the National Bureau of Standards), then later modified to its present design (Anderson *et al.*, 1978, Anderson and Alarie, 1978). This method is based on a method for studying the sensory irritation response of rodents (Alarie *et al.*, 1974).

The decomposition source is a conductively heated vertical tube furnace controlled by a linear programmer to increase temperature at 20°C per minute. The sample is held in a crucible in the furnace, with the crucible being on a Stanton Redcraft recording thermobalance. The furnace is connected to the exposure chamber by tubing which is cooled to reduce heat stress in the animals.

The 2.3 L glass chamber, shown in Figure 5-8, is a glass cylinder with parts fitted for the nose-only exposure of four mice at a time. The system is an open or dynamic one air flows in a single pass through the furnace and chamber and out of the system at 20 L/minute. Four male Swiss Webster mice are exposed for thirty minutes in this system. During exposure, everything but their noses is mounted in a plethysmograph - a sensitive device designed to monitor respiration by measuring changes in air pressure. In this system, changes in respiration rates and volumes are utilized to determine if sensory or pulmonary irritation (which are, in this usage, held to be equivalent to

Pittsburgh Mouse Exposure Chamber

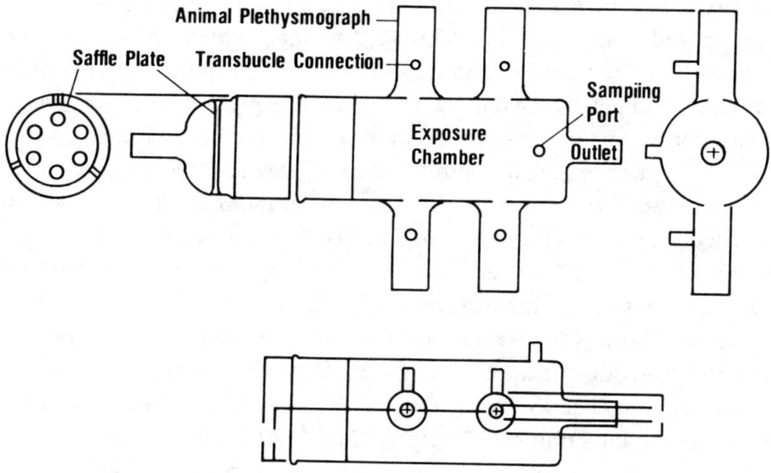

Figure 5.8

incapacitation) are induced. Animals are not held for observation after exposure.

Analytical support consists of GC monitoring for O_2, CO, and CO_2.

The protocol used here is based on the premise that a sample will, under these conditions, pass through each of the possible decomposition temperatures. Times to incapacitation and a concentration-lethality curve are determined.

Each material is evaluated for the toxic effects of its combustion products by the use of five toxicity indices or end points. Sensory irritation is evaluated in the same manner as in the previous method. A concentration-response relationship for sensory irritation and an RD_{50} value are obtained by measuring the average maximal decrease in respiratory rate within 10 minutes of the initiation of exposure to the smoke produced by a series of increasing sample weights. According to Alarie and his co-workers, the RD_{50} concentration in mice is equivalent to an intolerable level of sensory irritation in humans and is predicted to be incapacitating within 3 to 5 minutes (Alarie, 1973; Barrow *et al.*, 1978; Kane *et al.*, 1979).

A secondary toxicity end point used in this test method is the stress index (SI), also referred to as the sensory irritation stress index (SISI). According to Alarie and his co-workers, sensory irritants evoke a series of reflex responses, apart from the decrease in respiratory rate to compensate for the reduced breathing and apneic periods. The adjustments, which include changes in heart rate, blood pressure and blood vessel size, are representative of the total stress on the animal. The intensity of this stress is proportional to the decrease in respiratory rate, but the exact relationship has not been defined (Barrow *et al.*, 1979). A mathematical approximation of this relationship was proposed by the investigators for calculation of a stress index for the decomposition products of materials (Barrow *et al.*, 1979). This index takes into consideration the rate of onset and recovery for the respiratory effect as well as the degree of depression of the respiratory rate.

Concentration-response relationships may be obtained by graphing SI values as a function of sample size. From these relationships, the sample size of a material associated with an SI value of 100 is selected for comparison of potency among materials and is termed the SI100 value.

For screening purposes, investigators desire a simple test which will reveal materials having unusually high toxicity. For this test, they propose the use of only the lethality end point of their complete protocol. The LC_{50}, defined by them as the sample weight which causes death of half of the exposed animals within a fixed time, is selected as the lethality end point. Exposure of the four mice in plethysmographs, with heads extended into the chamber, begins with decomposition of the sample, as evidenced by weight loss, and continues for either 10 or 30 minutes. A 5 minute recovery period follows 10 minute exposures and a 10 minute recovery period follows 30-minute exposures. Lethality is determined by recordings of respiration and is defined as apnea (absence of breathing) persisting for 17 seconds. Any animal which dies during the exposure or the recovery period is counted in the LC_{50} calculation. Dose-response lethality curves of materials are prepared by graphing percent mortality as a function of the logarithm of sample weight and a classification system to rate the acute lethal potential of each material on the basis of its LC_{50} value is proposed. It is arrived at by placing the LC_{50} value of Douglas fir (taken as the reference material) at the middle of an order of magnitude on a logarithmic scale. Materials with LC_{50} values within this order of magnitude are classified "as toxic as wood." Those materials with LC_{50} values within one order of magnitude lower than this are classified "more toxic than wood"; similarly, materials with LC_{50} values within one order of magnitude higher than this are designated "less toxic than wood." The LC_{50} values determined by this test method were incorporated into a formula developed by Alarie and his co-workers for comparing the toxicological hazard of materials (Anderson and Alarie, 1978). Other variables used to calculate an acute lethal hazard (ALH) of a material are the temperature at 1-percent weight loss and the density of the material. The initial decomposition temperature was predicted to have an inverse relationship and the material density a direct relationship with the acute lethal hazard. Thus, a material which requires a high temperature for decomposition would tend to have a lower ALH value. Also, if the density of this material were low, these investigators made the assumption that less weight of material would be used and the hazard (ALH) would be less. Additionally, when comparing materials intended for the same structural function, other variables to define the amount of material used are incorporated into the formula. For example, for materials intended for insulation, the thermal conductivity (K) was used in addition to the density because both parameters define the total amount of material used for a particular insulating function. The thermal conductivity was predicted to have a direct relationship with the ALH. Under this assumption, a material with high thermal conductivity would be a poor insulator and require greater quantities, thereby resulting in a larger ALH value. Thus, the ALH for comparing the acute toxicological hazard of insulating materials was expressed as:

$$ALH = \frac{K \cdot D}{T \cdot LC_{50}}$$

where K is the thermal conductivity, D is the density, T is the temperature at which the material begins to decompose and LC_{50} is the sample weight lethal to half of the animals.

There are a number of strong objections to this system. The single pass nature of exposure permits the animals to conceivably avoid exposure to some extremely irritating decomposition stream components by holding their breath. No provision is made to avoid volatiles or particulates being condensed out prior to reaching the animals; and the assumption that the respiratory irritation model is in all cases equivalent to incapacitation models is not accepted by everyone. Additionally, materials with the same intended use, but with different thermal stabilities, are not evaluated under the same temperature conditions. There is also some question regarding the use of the RD_{50} value as an indicator of the incapacitating concentration of sensory irritants in man. The results of a study by Potts and Lederer (1978) do not support the predictions of Alarie and his co workers that the RD_{50} concentration of thermal decomposition products of a material would be rapidly incapacitating to man. In this study, mice and human volunteers were simultaneously exposed in a room to the decomposition products of a quantity of red oak calculated to produce an RD_{50} sensory irritation level for mice. Although the mice exhibited a greater than 50-percent decrease in respiratory rate, the human subjects complained of no ill effects, did not exhibit a greater than 50-percent decrease in respiratory rate and were definitely not incapacitated. However, strain differences in sensory irritation response by mice do exist and Potts and Lederer did not use the same strain of mouse as Alarie.

Advantages of the Pittsburgh method are that it provides the following: 1) an in-depth evaluation of toxic effects of the thermal decomposition products of a material by the use of five distinct end points; 2) determination of an LT_{50} value; 3) quantification of sensory irritation by determination of dose-response relationships and RD_{50} values; 4) determination of the asphyxiant range; and 5) detailed information relevant to the rates of thermal decomposition of a material and its release of toxicants. The method is also unique in its standard usage of ramped heating of the tube furnace (although this heating mode is an alternative procedure in the USF and improved FAA methods), and in the quantification of stress and the histopathological examination of exposed animals.

Federal Aviation Agency System

This test system was developed from 1974 on by a group lead by C. R. Crane (Crane *et al.*, 1977) at the FAA facility in Oklahoma city. The decomposition device utilized is a Lindberg conductive tube furnace, preheated to 600°C (± 1°C) before the sample (in a quartz tube) is placed in the furnace and the system is sealed. The system as diagrammed in Figures 5-9 and 5-10, is a closed loop "static" system.

Analytical monitoring consists of periodic determinations of HCN, CO, O_2 and CO_2 by two gas chromatographs.

For each trial, three male Sprague Dawley rats are placed in a motor driver activity wheel, the wheel being in a plexiglass box with a volume of 12.6 L. Exposure is for thirty minutes, but the furnace is connected for only ten minutes of this period. Both time-to-death and time-to-incapacitation are determined by observation. Surviving animals are watched for fourteen days post exposure.

Samples which consist of a low density material are cut into 1 cm cubes before being placed in the furnace. The protocol calls for determination only at different concentrations (the concentrations being determined by calculat-

FAA Exposure System

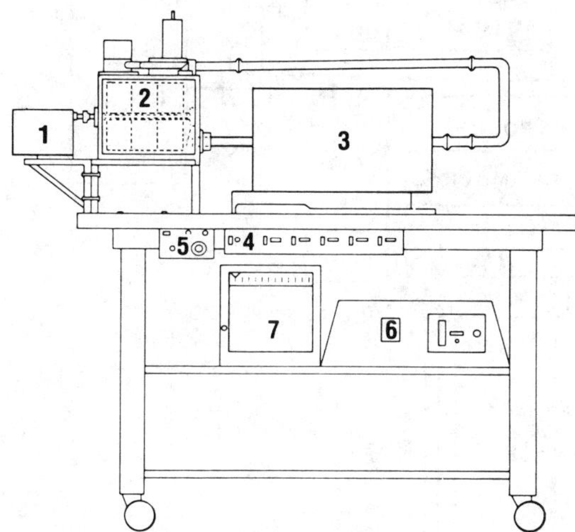

1. Motor to drive activity wheel
2. Animal exposure chamber
3. Tube Furnace
4. Electrical outlet strip
5. Recirculating fan controller
6. Furnace temperature controller
7. Chamber temperature recorder

Figure 5.9

ing a nominal concentration based on the weight of sample volatilized) achieved by heating the sample to 600°C. This is adequate for most materials, but: (1) there are some which only decompose at higher temperatures and (2) higher temperatures are achieved in a real fire. And decomposition, in many materials, progresses over a period exceeding ten minutes. The relatively small number of animals utilized per trial also serves to reduce somewhat the sensitivity of the system.

There were several marked limitations to the use of the first FAA test system for the evaluation of materials. The 10-minute decomposition time at 600°C might not produce complete decomposition of some materials. Recirculation of the thermal decomposition products between the furnace and chamber could possibly cause further degradation of these products and/or loss of products by reaction or absorption. The test method did not quantify toxic potency by determination of dose-response relationships and LC_{50}

FAA Exposure System

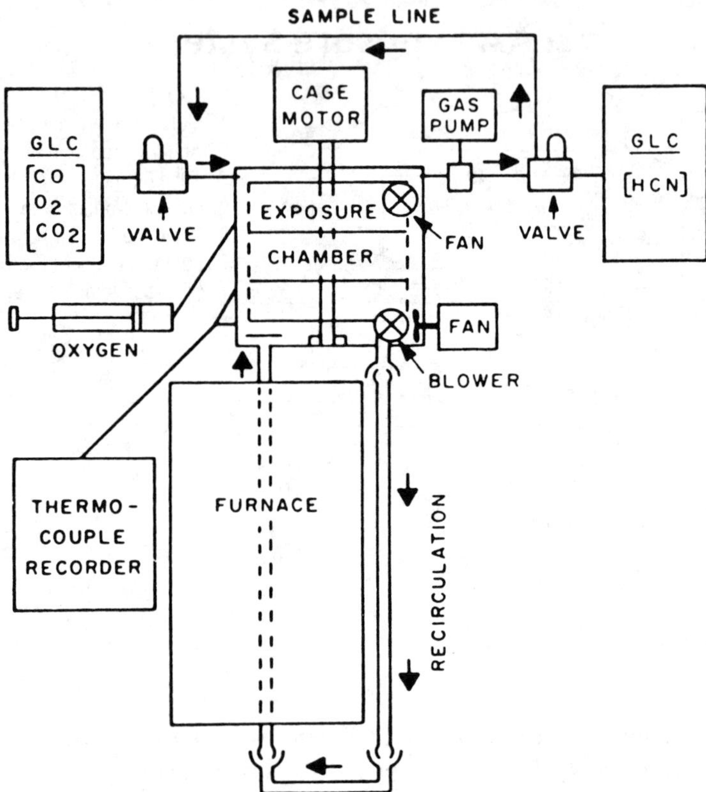

Figure 5.10

values. Finally, determination of time-to-incapacitation and time-to-death by observation might be obscured in dense smoke-containing atmospheres. This latter possibility was eliminated by modification of the rotating wheel to enable electronic monitoring of animal responses. However, it was reported that rotation of the wheel cannot be stopped until the last animal is incapacitated and, consequently, T_ds occurring prior to the final T_i cannot be recorded (Spurgeon, et at., 1979).

Most of these limitations have been circumvented by modifications to the system which now allow fixed or ramped heating to 800°C and variation of sample size to quantify toxic potency. However, the improved test system also is a recirculating system, with the inherent potential limitations described. Advantages of the improved system are 1) its versatility in providing fixed or ramped heating and nonflaming or flaming combustion, and 2) its provision of time-to-incapacitation and time-to-death data in addition to the quantification of toxic potency.

DIN Method

This test method was developed in the Federal Republic of Germany and has been used by a number of German investigators, including G. Kimmerle, H. T. Hoffmann, H. Oettel, H. J. Klimisch, H. Hollander and others (Kimmerle and Praeger, 1980; Klimisch *et al.*, 1980; Hoffmann and Sand, 1976; Klimisch *et al.*, 1980). The method was developed in response to the German Commission of Standards desire for a standardized test method for generating combustion products in toxicity studies of smoke produced by materials. The standardized method has been designated DIN 53 436.

The basic components of test protocols used in conjunction with the DIN method consist of the following (Kemmerle and Prager, 1980).

Combustion Apparatus

The combustion apparatus and procedure, are used for generation of thermal decomposition products. The apparatus consists of a 1300-mm long quartz combustion tube with an outside diameter of 39 mm and a moveable, circular electric oven which tightly encloses a section of the tube. The oven moves along the tube at a rate of 10 mm per minute. The procedure permits variable thermal exposure of the test material in a controlled air stream. Air enters the tube at a rate of approximately 100 L per hour in the direction opposite to the movement of the oven. The test specimens, cut so as to have an equal volume or weight per unit length, are heated in the quartz tube at different constant furnace temperatures between 200°C and 600°C. The mixture of air and combustion products from the tube is cooled and further diluted by a secondary stream of air, controlled by a flowmeter, prior to passage into and through an animal exposure chamber.

Exposure Chamber

The animal chamber is operated in the dynamic mode for exposure of rats to the diluted combustion mixture. A variety of animal exposure chambers of different sizes and configurations have been used with the DIN apparatus to provide both head-only and whole-body exposures of animals, but usually the latter. At least 5, but usually 20, animals are used in each test. In most tests, the duration of exposure is 30 minutes.

Biological End Points

Toxicological measurements consist of mortality as the primary end point, exposure and post-exposure observations of toxic signs, animal body weight determinations and gross pathological examinations.

Atmosphere Analysis

Chemical analyses, consisting of continuous monitoring of CO, CO_2, and O_2, periodic measurements of HCN or other selected gases, and measurements of blood COHb in animals at the termination of exposure are performed.

The DIN method may be used to assess the relative toxicities of combustion products of a material under different temperatures or to compare the relative toxicities of combustion atmospheres produced by different materials. Toxicity may be expressed in terms of mortality as the ratio of the number of dead animals to the number of animals exposed. Mortality values may also be related to the total air flow and to either the volume of the specimen heated during the 30-minute period or to the mass or mass loss of the specimen. A dose-response relationship and LC_{50} may also be determined for each material by varying the volume of diluting air in order to vary the concentration of the combustion mixture to which the animals are exposed.

A defined critical temperature (T_C) has been used in a number of studies with the DIN method to assess the relative toxicity of the pyrolysis products of plastics. In addition to the critical temperature, Kimmerle and Prager (1980) have used a number of other parameters to compare relative toxicities.

One of the advantages of the DIN method is the reported excellent reproducibility of the method, when used by different laboratories to generate combustion atmospheres for acute toxicity studies. This reproducibility was evaluated in recent studies by Klimisch and co-workers (Klimisch *et al.* , 1980; Klimisch, 1980). Three laboratories participated in the evaluation which consisted of 1) measurements of toxicity in animals exposed to thermal decomposition products 2) measurements of carbon monoxide and carbon dioxide in inhalation chambers; and 3) measurements of total volatile organic substances in inhalation chambers. The results for each of the three parameters by the three laboratories demonstrated clearly that reproducible results may be obtained in different laboratories using the DIN method. Another advantage of this method is its versatility in allowing both whole-body and head-only exposures of test animals.

There are two features of the DIN combustion device and technique which are unique. One is that the tube furnace moves along the test material at a fixed rate, thermally decomposing a constant quantity of material per unit time. This enables the generation of a defined concentration of thermal products that can be kept steady during the whole period of exposure. The second is that concentrations of combustion products can be obtained for determining dose-response relationships by diluting the combustion mixture while maintaining the same proportion of individual components.

One limitation of the DIN method is that furnace temperatures are limited to 600°C, although temperatures in real fires can exceed this temperature. Additionally, time to-incapacitation is not measured.

Southwest Research Institute (SWRI) system

The SWRI system is essentially a modification of the NBS system. The changes consist primarily of a more complete analytical package being utilized. A Beckman diesel analyzer serves to continuously measure and record levels of CO, CO_2, O_2, NO/NO_X, and total hydrocarbons, while a gas chromatograph/mass spectrophotometer arrangement is available for specialty gases. The system is otherwise just like the NBS system and the protocol utilized is the same.

Utah Flamability Research Center System

Developed under contract to the NBS, this system (Petajan, 1977) is the product of a team at the FRC in salt Lake city. It uses the Pott's pot as a decomposition device, with the furnace having its open upper face flush with the bottom of a 64 liter, octagonal polymethylmethacrylate exposure chamber. The system is a closed, "static" system, with a flexible bag attached to allow for expansion of total gas volume.

Four male Long Evans rats receive 30 minute, head only, exposure to concentrations (calculated based on weight volatilized) of test materials, with fourteen days of post-exposure observation for survivors. The leg flexion model is utilized here to evaluate incapacitation. An extensive analytical support package based on a GC mass spectrophotometric system is included and a wide range of blood parameters (total hemoglobin, COHb, O_2Hb, and blood gases) is determined. The protocol utilized requires concentration response curves for each of two sample decomposition modes (just below auto ignition and just above auto ignition). The small number of animals utilized does serve to reduce sensitivity to some degree.

University of Michigan System

Developed by a group lead by Boetner and Hartung under a Society of Plastics Industry contract from 1975 to 1978, several configurations of systems were evaluated by this group, but the one finally adapted is described in Hartung *et al.* (1977). A radiant panel heater is used to decompose samples in this closed system.

Analytical support consists of infrared measurement of carbon monoxide and, when needed, colorimetric or gas chromatographic analysis of special interest gases.

The exposure system is a 370 liter plexiglass cube with a rotarod mounted in it. The entire system is static with recirculation within the chamber.

Each trial consists of a thirty minute exposure of four male Sprague Dawley rats, with inability to stay on the rotarod being evaluated as incapacitation. Concentration lethality and time to incapacitation curves are determined at 10° below and above the auto ignition temperature.

University of San Francisco System

C. J. Hilado developed this system under contract for NASA. It has been extensively described in the literature (Hilado *et al.*, 1976; Hilado and Crane, 1977)

Decomposition is achieved by a conductively heated tube furnace attached by a non-insulated tube to a polymethylmethacrylate exposure chamber which is 4.2 liters in volume and hemispherical in shape (see Figure 5-11). Analytical support consists of continuous oxygen and temperature monitoring (but without continuous recording). Four male Swiss albino rats are free to move about the exposure chamber during the course of the thirty minute test. An observer judges both incapacitation and death of animals, recording a time to each of these endpoints. The system is a dynamic, open one; decomposition products make a single passage through the chamber and out of the system.

There are a number of serious problems with this system. The control over decomposition conditions is extremely limited. Animals have ample opportunity to avoid breathing toxic chemical species which are present for only short periods. Finally, the analytical measurements in this system are just too minimal to allow any real evaluation of relationships to toxicant concentration to be made.

The USF test method has been recommended by Hilado as a simple, economical and rapid screening test. However, the method provides only limited toxicological and analytical data. It is unique in its flexibility to

USF/NASA Exposure Chamber

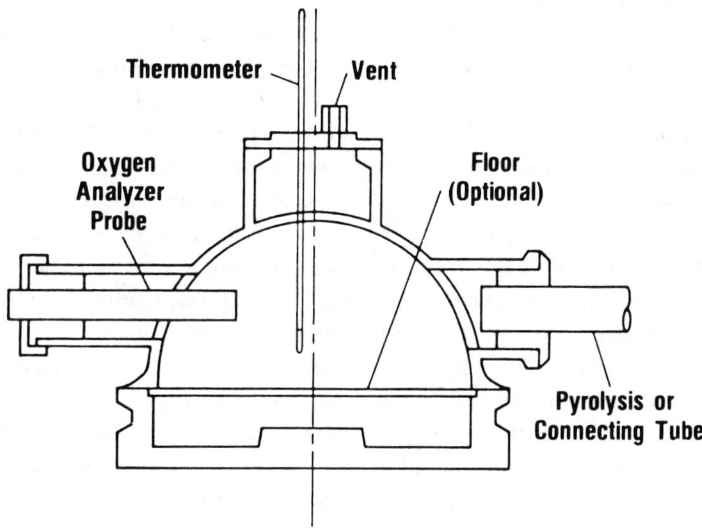

Figure 5.11

operate the tube furnace under either fixed or ramped temperature conditions and in either a static or dynamic air flow mode.

One of the limitations of the USF test protocol is that it does not quantify toxic potency of the decomposition products of a material. As a consequence, the USF method is the only system which did not identify polytetrafluorothylene (PTFE) as producing highly toxic smoke.

The extreme toxicity of the decomposition products of this material has been demonstrated by several other test methods. The modified protocol, the Dome Chamber Toxicity Test method, does provide for determination of dose-response lethality relationships and LC_{50} and LT_{50} values.

The USF method evaluates incapacitation and death by visual, and therefore subjective, measurements. Although these seem to be reproducible and require no animal training, more objective end points would be preferable. The use of three end points (staggering, convulsions and collapse) by Hilado to provide insight into the nature of the toxic species is also subjective, is not supported by any explanation of mechanisms of action, and has not been confirmed by any other researchers. The fact that the animals are allowed to run free is an advantage in setup costs and maintenance, but may be a disadvantage because the rodents can huddle together and "protect" one another from irritant gases.

Identification of the cause of incapacitation or death is an important aspect of combustion toxicology which has been emphasized in studies by numerous authors. This capability is limited in the USF methodology, primarily because of the small sample size and small chamber volume (which preclude extracting many samples for analysis). In his more recent studies, however, Hilado made measurements of carbon monoxide in an effort to relate the cause of death to CO concentration-time (C•t) relationships. Although these data appear to be reasonable, conclusions based on a single point measurement and an assumption of the pattern of CO evolution with time could easily be in error by 50 percent or more in the C•t product. More extensive studies of the CO evolution from various materials should be conducted to verify the conclusions reached. Another unknown factor in the USF methods which influences the combustion atmosphere is the length of travel of products down the combustion tube without the aid of air flow. It would seem that some of the combustion products could remain in the tube long after they were released.

SUMMARY

These systems still need critical evaluation in terms of being (as discussed earlier) a primary (pass/fail), secondary (ranking), or tertiary (total evaluation) level system. Most, if not all, of these systems can operate as primary level systems - they will identify those materials that have a high probability of generating extremely toxic degradation materials. Several of the systems (with some weaknesses) can act as secondary level evaluations while the NBS, CMIR, SWIR, FAA, and University of Michigan systems can be utilized to generate comparative hazard data. What is needed for a tertiary level system will be discussed later. One universal weakness in all existing systems is that they do not quantitate and evaluate smokes and other aerosols. And they all allow at least some possibility that the hazards associated with aerosols will not be adequately evaluated.

REFERENCES

Alarie, Y. (1966) Irritating Properties of Airborne Materials to the Upper Respiratory Tract. *Arch. Environ. Health* **13**: 433.

Alarie, Y. (1973) Sensory Irritation by Airborne Chemicals. *CRC Crit. Rev. Toxicol.* **2**: 299.

Alarie, Y. (1981) Dose-response Analysis in Animal Studies: Prediction of Human Responses. *Environ. Health Perspect.* **42**: 9.

Alaire, Y. (1985) The toxicity of Smoke from Polymeric Materials during Thermal Decomposition. *Ann. Rev. Pharmacol. Toxicol.* **25**: 35.

Alarie, Y. and Anderson, R. C. (1981) Toxicologic Classification of Thermal Decomposition Products of Synthetic and Natural Polymers. *Toxicol. Appl. Pharmacol.* **57**: 181.

Alarie, Y and Barrow, C. S. (1977) *Toxicity of Plastic Combustion Products. Toxicologic to Assess the Relative Hazards of Thermal Decomposition Products from Polymeric Materials.* National Technical Information Service Publication PB267 23 ; 286 pp.

Alarie, Y., Lin, C. K. and Geary, D. L. *Sensory Irritation Evoked by Plastic Decomposition Products.* Presented at the American Industrial Hygiene Conference Miami, Florida (May 15, 1974).

Alexeeff, G. and Packham, S. C. (1984) Evaluation of sMoke Toxicity Using Concentration-time Products. *J. Fire Sci.* **2**: 362.

Anderson, R. C. and Alarie, Y. C., (1978) An Attempt to Translate Toxicity of Polymer Thermal Decomposition Products Into a Toxicological Hazard index and Discussion on the Approaches Selected. *Journal of Combustion Toxicology.* **5**: 476.

Anderson, R. C. and Alarie, Y. C. (1978) Screening Procedure to Recognize "Supertoxic" Decomposition Products From Polymeric Materials Under Thermal Stress. *Journal of Combustion Toxicology* **5**: 54-63.

Anderson, R. C., Dierdorf, J., Stock, M. F., Matijak, M., Sawin, R. and Alaire, Y. (1978) Use of Experimental Materials to Assess Toxicological Tests for Rating Polymeric Materials Under Thermal Stress. *Fire Mater* **2**: 136-140.

Autian, J. (1970) Toxicologic Aspects of Flamability and Combustion of Polymeric Materials. *J. Fire Flam.* **1**: 239.

Barrow, C. S., Alarie, Y., and Stock, M. F. (1976) Sensory Irritation Evoked by the Thermal Decomposition Products of Plasticized Polyvinyl Chloride. *Fire Materials* **1**: 147.

Barrow, C. S., Alarie, Y., and Stock, M. F. (1978) Sensory Irritation and Incapacitation Evoked by Thermal Decomposition Products of Polymers and Comparisons with Known Sensory Irritants. *Arch. Environ. Health* **33**: 79.

Barrow, C. S., Lucia, H. and Alarie, Y. C. (1979) A Comparison of the Acute Inhalation Toxicity of Hydrogen Chloride Versus the Thermal Decomposition Products of Polyvinylchloride. *Journal of Combustion Toxicology,* **6**: 3-12.

Barrow, C. S., Lucia, H., Stock, M. F. and Alarie, Y. (1979) Development of Methodologies to Assess the Relative Hazards from Thermal Decomposition Products of Polymeric Materials. *J. Amer. Ind. Hyg. Assoc.* **40**: 408-423.

Birky M. M. (1976) Philosophy of Testing for Assessment of Toxicological Aspects of Fire Exposure. *J. Combust. Toxicol.* **3**: 5-23.

Birky, M. M. (1977) *Hazard Characteristics of Combustion Products in Fires: the State-of-the-art Review.* NBSIR 77-1234, Washington, DC, National Bureau of Standards.

Birky, M. M., Paabo, M. and Brown, J. E. (1980) Correlation of Autopsy Data and Materials Involved in the Tennessee Jail Fire. *J. Fire Saf.* **2**: 17-22.

Boettner, E. A. and Hartung, R. (1978) *The Analysis and Toxicity of the Combustion Products of Natural and Synthetic Materials.* Final Report to the Society of the Plastics Industry, Inc. and the Manufacturing Chemists Association.

Clarke, III, F. B., Benjamin, I. A., and Clayton, J. W. (1982) An Analysis of Current Knowledge in Toxicity of the Products of Combustion. **In**: Report to Committee on the Toxicity of the Products of Combustion of the National Fire Protection Association. Morgan Technical Library, Batterymarch Park, Quincy, MA.

Crane, C. R., Sanders, D. C., Endecott, B. R., Abbott, J. K., and Smith, P. W. (1977) *Inhalation Toxicology: I. Design of a Small-Animal Test System. II. Determination of the Relative Toxic Hazards of 75 Aircraft Cabin Materials*. Report No. FAA-AM-77-9, Department of Transportation, Federal Aviation Administration. Washington, DC, Office of Aviation medicine.

Crane, C. R., Sanders, D. C., Endecott, B. R., and Abbott, J. K. (1985) *Inhalation Toxicology IV. Times to Incapacitation and Death for Rats Exposed Continuously to Atmospheric Hydrogen Chloride Gas*. FAA-AM-85-4, Civil Aeromedical institute, Oklahoma City, OK, Federal Aviation Administration.

Dilley, J. V., Martin, S. B., McKee, R., and Pryor, G. T. (1979) A Smoke Toxicity Methodology. *J. Comb. Tox.* **6**: 20.

Finney, D. J. (1971) *Probit Analysis, 3rd edition*. Cambridge University Press.

Edgerly, P., Rogowski, Z., and Wooley, W. (1982) A Study of the Combustion Processes of the Pott's Pott Method and DIN Method Using Wood, Polypropylene and Polymethyl methacrylate. *Proceedings of Interflam Conference 1982, London, England.*

Gad, S. C. (1988) *Handbook of Product Safety Evaluation*. Marcel Dekker, New York.

Gad, S. C. (1988) An Approach to the Design and Analysis of Screening Data in Toxicology. *Journal of the American College of Toxicology* **7**: 127-138.

Gad, S. C., and Smith, A. C. (1984) Influence of Heating Rates on the Toxicity of Evolved Combustion Products: Results and a System for Research. *J. Fire Science* **1**: 465-479.

Gad, S. C. and Weil, C. S. (1988) *Statistics and Experimental Design for Toxicologists*. Telford Press, Caldwell, NJ.

Gaume, J. G. (1980) Instrumental Animal Systems for Toxic Assessment of Materials. *J. Comb. Tox.* **7**: 124.

Gaume, J. G., Reibold, R. C., and Spieth, H. H. (1981) Initial Tests of the Combined ECG/Ti Animal Sytems Using Carbon Monoxide Exposure. *J. Comb. Tox.* **8**: 125.

Hartung, R., Ball, G. L., Boettner, E. A., Rosenbaum, R., and Hollingsworth, Z. R. (1977) The Performance of Rats on a Rotarod During Exposure to Combustion Products of Rigid Polyurethane Foams and wood. *J. Combust. Toxicol.* **4**: 506-522.

Hartzell, G. E., Priest, D. N., and Switzer, W. G. (1985) Modeling of Toxicological Effects of Fire Gases: ii. Mathematical Modeling of Intoxication of Rats by Carbon Monoxide and Hydrogen Cyanide. *J. Fire Sci.* **3**: 115.

Hartzell, G. E., Stacy, H. W., Switzer, W. G., Priest, D. N., and Packham, S. C. (1985) Modeling of Toxicological Effects of Fire Gases: IV. Intoxication of Rats by Carbon Monoxide in the Presence of an Irritant. *J. Fire Sci.* **3**: 263.

Hilado, C. J. and Crane, C. R. (1977) Comparison of Results with the USF/NASA and FAA/CAMI Toxicity Screening Test Methods. *J. Combust. Toxicol.* **4**: 56-60.

Hilado, C. J., Saxton, G. L., Kourtides, D. A., Parker, J. A., and Gilwee, W. J. (1976) Relative Toxicity of Pyrolysis Products of Some Cellular Polymers. *J. Comb. Tox.* **3**: 259.

Hilado, C. J., Slattengren, C. L., and Furst, A. (1976) Relative Toxicity of Pyrolysis Products of Some Synthetic Polymers. *J. Comb. Tox.* **3**: 270.

Hilado, C. J., Solis, A. N., Marcussen, W. H., and Furst, A. (1976) Effect of Temperature and Heating Rates on Apparent Lethal Concentrations of Pyrolysis Products. *J. Combust. Toxicol.* **3**: 381-392.

Hofmann, H. Jr. and Oettel, H. (1969) Relative Toxicity of Plastics Combustion Products, Particularly Rigid Polystyrene Foam. *Kunstst Rundsch* **15**: 261-8.

Hoffmann, H. and Sand, H. E., (1976) Evaluation of the Acute Toxicity Hazard of Combustion Products Given Off by Organic materials. **In**: *Proceedings: International Symposium on Toxicity and Physiology of Combustion Products*. National Academy of Sciences Committee on Fire Research, National Research Council, Salt Lake City, Utah (March 22-26, 1976).

Kane, L. E., Barrow, C. S. and Alarie, Y., (1979) A Short-Term Test to Predict Acceptable Levels of Exposure to Airborne Sensory Irritants. *Amer. Ind. Hyg. Assoc. J.* **40**: 207-229.

Kaplan, H. L., Anzueto, A., Switzer, W. G., and Hinderer, R. K. (1986) Respiratory Effects of Hydrogen Chloride in the Baboon. *The Toxicologist* **6**: 52.

Kaplan, H. L., Grand, A. F., and Hartzell, G. E. (1983) *Combustion Toxicology: Principles and Test Methods*. Lancaster, PA, Technomic Publishing Company, Inc., pp. 1-174.

Kaplan, H. L., Grand, A. F., Switzer, W. G., and Gad, S. C. (1984) Acute Inhalation Toxicity of the Smoke Produced by Five Halogenated Polymers. *J. Fire Sciences* **2**: 153-172.

Kaplan, H. L., Grand A. F., Switzer, W. G., Mitchell, D. S., Rogers, W. R., and Hartzell, G. E. (1985) Effects of Combustion Gases on Escape Performance of the Baboon and the Rat. *J. Fire Sci.* **3**: 228.

Kaplan, H. L. and Hartzell, G. E. (1984) Modeling of Toxicological Effects of Fire Gases: I. Incapacitating Effects of Narcotic Fire Gases. *J. Fire Sci.* **2**: 286.

Kimmerle, G. (1974) Aspects and Methodology for the Evaluation of Toxicological Parameters During Fire Exposure. *J. Fire Flam./Comb. Tox.* **1**: 4.

Kimmerle, G. and Prager, F. C., (1980) The Relative Toxicity of Pyrolysis Products. Part II. Polyisocyanate Based Foam Materials. *Journal of Combustion Toxicology* **7**: 54.

Kishitani, K. and Nakamura, K. (1974) Toxicities of Combustion Products. *J. Fire Flam. Combust. Toxicol.* **Suppl. 1**: 104-123.

Klimisch, H., (1980) Generation of Constant Concentrations of Thermal Decomposition Products in Inhalation Chambers. A Comparative Study with a Method According to DIN 53 436. II. Measurement of Concentrations of Total Volatile Organic Substances in Inhalation Chambers. *Journal of Combustion Toxicology* **7**: 257.

Klimisch, H., Hollander, H. W. M. and Thyssen, J., (1980) Generation of Constant Concentrations of Thermal Decomposition Products in Inhalation Chambers. A Comparative Study with a Method According to DIN 53 436. I. Measurement of Carbon Monoxide and Carbon Dioxide in Inhalation Chambers. *Journal of Combustion Toxicology* **7**: 243.

Klimisch, H., Hollander, H.w.m. and Thyssen, J., (1980) Comparative Measurements of the Toxicity of Laboratory Animals of Products of Thermal Decomposition Generated by the Method of DIN 53 436. *Journal of Combustion Toxicology* **7**: 209.

Levin B. C., Fowell, A. J., Birky, M. M., Paabo, M., Stolte, A., and Malelz, D. (1982) *Further Development of a Test Method for the Assessment of the Acute Inhalation Toxicity of Combustion Products*. NBSIR 82-2352, Washington, DC, National Bureau of Standards.

MacFarland, H. N. (1977) Respiratory Toxicology. **In**: *Essays in Toxicology* **7**: 141. Edited by W. Hayes, Jr. New York, Academic Press.

NRC Committee on Fire Toxicology (1977) *Fire Toxicology: methods for evaluation of toxicity of pyrolysis and combustion products*. Report No. 2, Natl. Acad. Sci., Washington, D.C. p.34.

NRC Committee on Fire Safety Aspects of Polymeric Materials (1978). *Smoke and Toxicity (Combustion Toxicology of Polymers)* Vol. 3, Publication NMAB 318-3, Nat. Acad. Sci., Westport, CT: Technomic Publishing Co.; p. 55.

National Bureau of Standards. (1982) *Further Development of a Test Method for the Assessment of the Acute inhalation Toxicity of Combustion Products*. NBSIR 82-2532 (June 1982).

National Materials Advisory Board. (1978). *Smoke and Toxicity (Combustion Toxicology of Polymers)*, Vol. 3. Publication No. NMAB 318-3, Washington, DC, National Academy of Sciences, pp. 1-55.

N.Y. Proposed Article 15, Part 1120 *Combustion Toxicity Testing. (1985). Uniform Fire Prevention and Building Code Council*. Albany, New York: State of New York, Department of State, 12231.

Packham, S. C., Jeppsen, R. B., McCandless, J. B., Blank, T. L., and Petajan, J. H. (1978). The Toxicologic Contribution of Carbon Monoxide as a Component of Wood Smoke. *J. Comb. Tox.* **5**: 11.

Petajan, J. H. (1977) An Approach to the Toxicology of Combustion Products of Materials. *Environ. Health Perspect.* **17**: 65-73.

Potts, W. J. and Lederer, T. S. (1977) A Method for Comparative Testing of Smoke Toxicity. *J. Comb. Toxicol.* **4**: 114-162.

Potts, W. J. and Lederer, T. S., (1978) Some Limitations in the Use of Sensory Irritation Method as an End-Point in Measurement of Smoke Toxicity. *Journal of Combustion Toxicology* **5**: 182-195.

Rodkey, F. L. and Jenkins, L. J. (1979) Toxicity of Shipboard Fire Combustion Products. *Toxicol. Res. Proj. Direct.* 4.

Spurgeon, J. (1978) *The Correlation of Animal Response Data with the Yields of Selected Thermal Decomposition Products for Typical Aircraft Interior Materials.* FAA-NA-78-45, Civil Aeromedical Institute. Oklahoma City, OK, Federal Aviation Administration.

Spurgeon, J. C., Filipczak, R. A., Feher, R. E. and Sternik, S. J., (1979) A Procedure of Electronically Monitoring Animal Response Parameters Using the Rotating wheel. *Journal of Combustion Toxicology* **6**: 198.

Sumi, K. and Tsuchiya, Y. (1975) Toxicity of Decomposition Products. *JFF/Comb. Tox.* **2**: 213.

Tsuchiya, Y. (1981). Dynamic Toxicity Factor-evaluating Fire Gas Toxicity. *J. Comb. Tox.* **8**: 187.

Waritz, R. S. (1975) Industrial Approach to Evaluation of Pyrolysis and Combustion Hazards. *Environ. Health Perspect.* **11**: 197-202.

Wright, P. L. and Adams, C. H. (1977) Toxicity of Combustion Products from Burning Polymers: Development and Evaluation of Methods. *Environ. Health Perspect.* **17**: 75-83.

Chapter 6

REGULATIONS CONCERNING SMOKE TOXICITY

Rosalind C. Anderson

RECENT CONCERNS

Fires which result in both multiple deaths and publicity are those which serve as the impetus for legislation, typically aimed at establishing new limits or guidelines to protect the public safety. A variety of regulatory approaches has been used although the vehicle tends to be the building code. Most frequently the approach has been to require an additional layer of fire protection such as sprinklers or early warning devices. In addition, there may be an identification of, and special limits devised for, high risk occupancies. On occasion, the response has been to identify a guilty product and to prohibit the use of the product or its component materials in the local jurisdiction.

In terms of regulation, toxicity, which is the recorded cause of the majority of fire deaths, is approached only obliquely. The logic has been that with no fire there is no fire death and that all is by definition " bad". At the same time that fire statistics linking fire death to toxic atmospheres became well publicized there was a growing conviction among fire fighters that there was a dramatic change in the atmospheres that they encountered on a daily basis. This perception, linked with economically motivated interests, activated a very aggressive conflict concerning the "new and greater" toxicity of fires.

Little evidence has been available to support or to dismiss the expectation that the increasing use of new materials in the built environment represented a parallel and ominous change in the safety of individuals exposed to fire effluents. The fear that we are enclosing ourselves by more and more toxic materials becomes reactivated as dramatic fires continue to occur.

An early approach which appears, at least in part, to have been intended to address the issue of smoke toxicity was the requirement for smoke density testing of some materials and products. In addition to providing an indication of visual obscuration there was an underlining and primarily informal hope that the results would provide an indication of potency. Certainly there is still a very common perception that dark smoke, or among the more sophisticated observers, green smoke, is the most toxic variety. Smoke density tests have proven to measure smoke *density*, as might have been predicted, and other avenues have been selected for the evaluation and limitation of smoke *toxicity*.

Although there has been little actual legislation, the toxicity of the fire atmosphere is a subject of public discussion in the United States following every major fatal fire. The discussion of fire toxicity has an urgency and conviction which places it with generic motherhood and public safety issues. The translation from a subject of public anger to an arena of scientific study and control has been difficult.

FORMAL PROVISIONS FOR CONTROL OF SMOKE TOXICITY

In the effort to limit the human toll a general approach has been to identify the good old days as a standard of excellence and to try to hold that line by prescribing performance similar to that of traditional products. A series of codes was developed which specified "no more toxic than wood or paper (the traditional materials) when burned under similar conditions." The difficulty in implementing this language as a regulation was greatly compounded by the absence of a commonly admired test procedure. This language also posed a difficulty in that it assumed wood to be a homogeneous material having a single predictable and desirable pattern of performance.

The failure to provide a clearly identified standard reference material, to specify a standardized test and a numerical passing value resulted in a provision which did not satisfy the traditional building code model in which levels of performance are specified by the regulatory agency with a stringency proportional to the perceived risk and level of concern.

These deficiencies in the wording and underlying assumptions of the early toxicity provisions of the model codes made these smoke toxicity regulations unworkable and resulted in the deletion of these requirements except in the cities of New York and Los Angeles. In the city of New York there has been, and still is, a requirement that products of special concern be shown to pass a toxicity test by comparison to a wood standard.

The next major civilian effort to control the use of building materials with respect to toxicity has been instituted by New York State as part of the revision of the state building code. This regulation is unusual in that it does not impose a passing standard for covered materials but instead institutes a filing requirement for specified products. Because an emphasis on toxicity with reduced attention to other fire properties was considered to represent a predictable and unprofitable interpretation of the new requirements the documents which manufacturers submit to the state must report toxicity as well as any other available fire test data. The manufacturers indicate on a separate confidential portion of the submission the chemical formulation of the product being registered.

Smoke toxicity is an issue which is being toyed with in a significant number of agencies. The Department of Defense issues specifications for purchasing which in the circulated draft versions often incorporate some smoke toxicity requirement. To a great extent these requirements do not survive to the last round of revisions, primarily because of the lack of a standardized method. Lack of agreement on the proper measurement parameters is a continued problem but possibly less difficult than the absolute insistence on a method which has survived a consensus standardization process. None of the available methods fill that need.

OTHER REGULATIONS CONCERNING TOXICITY OF SMOKE

In France, the Ministry of the Interior has instituted requirements which are satisfied by chemical analysis or by examination of the product formulation. The aim is to prevent the interior furnishings of a room from providing more than 5 grams of HCN or 25 grams of HCL per cubic meter during a fire.

A requirement for chemical analysis of smoke was instituted by the British Navy and adopted by the U.S. Navy for testing some materials. In NES 713 and the derivative tests, a weighed sample is burned inside a closed system. Draeger tube analysis of smoke components provides a semiquantitative indication of the levels of the gases present in the chamber. The gas concentrations are summed to provide a value which describes the mathematical potency of the test atmosphere.

Japanese regulations focus on the materials used on walls in fire exit corridors. A biological test has been developed in which the time to incapacitation is measured. To pass the test, the animal incapacitation time for the candidate product must be equal to or longer than that resulting from a standard luan board. Other products in Japan are not regulated for fire toxicity.

The remainder of this chapter will focus on the regulations in effect in the city and state of New York. A comparison of the requirements and effects of the New York City and New York State regulations provides an instructive contrast in two uses of the building code. Clearly, the intentions of these two approaches to material control are not the same. At first review it appears that state regulation, which requires filing of toxicity test data, can have no contribution to the safety of building inhabitants because of the absence of the demarcation between acceptable and unacceptable performance and because of the number of uncontrolled materials in the same potential fire environment.

Further study of the regulation and the manufacturers' response to it would suggest that a very effective philosophy of regulatory manipulation is at work and that the state requirement may have far more effect on the manufacturing community and the direction of new product development than the traditional performance test, as exemplified by the City.

NEW YORK STATE

The portion of the Uniform Fire Prevention Code which covers toxicity requirements for the state is 9 NYCRR 1120 Article 15 (Appendix A). The implementing regulations are found in 9NYCRR 439 (2).

Test And Test Laboratories

The test which has been selected by New York State was developed at the University of Pittsburgh. The Pittsburgh test (1) was selected in preference to the National Bureau of Standards Protocol (NBS), one of the few alternative, fully developed tests (4). This choice was based primarily on the broad applicability of the Pittsburgh test. The NBS test was found to provide results similar to those from the Pittsburgh test for many products but to be severely limited in the study of low density materials. The specific protocol required by the state for testing by means of the Pittsburgh test for combustion product toxicity is included as Appendix A.

This is a biological test in which male Swiss Webster mice are used as subjects. The animal test is considered to be more easily interpreted than the available chemical tests because the animals serve to integrate the various effects of the changing concentrations of multiple simultaneous chemical exposures. These effects are not easily modeled, chemical interactions being as yet difficult to predict.

The Pittsburgh test is classed as dynamic because smoke is moved in an air stream from the point of origin in the furnace to the animals, which are in a physically separate chamber. The test sample is decomposed by being heated in a furnace at a rate of 20°C over a 30 minute period. The furnace temperature at which the animal exposure is initiated (T_0) is determined by the weight loss of the sample. As a result there is a different temperature profile for the testing of different products. In addition, spontaneous flaming may occur during the test. At the end of the 30 minutes there is seldom organic material remaining. Mortality is determined following the thirty minute exposure and ten minutes post exposure observation period. The toxicity of the test sample is defined as the furnace load which would be statistically predicted to kill 50% of the test animals.

The reproducibility of the Pittsburgh test has not been demonstrated by a formal round robin study. However, a series of 40 replicate tests using nylon 6/6 has been conducted by two test laboratories with a total of 10 furnaces. The test results agreed within a factor of two. Greater divergence is typically noted with smokes containing more reactive chemicals, such as HCl.

There is some concern that the short animal observation time selected by the state in the official protocol results in an uneven assessment of smoke potency. Toxic agents, such as CO and HCN which kill very quickly, are more severely judged than smoke in which the principal toxic agent is a slow acting chemical. The example which is most frequently cited in this regard is HCl which typically causes deaths after a 4 to 48 hour latency. A longer post exposure observation period has resulted in lower test values for products releasing significant amounts of HCl but not for products releasing CO or HCN.

Laboratories are certified by the state to perform this testing. The certification procedure requires test laboratories to demonstrate that the laboratory is independent of any manufacturer as sponsor.

The laboratory must describe its experience with the required test procedure, name individuals performing the experiments and their training. The laboratory which wishes certification must also provide a copy of the protocol and quality assurance (QA) procedures used. The laboratories which are certified (May 1989) are Anderson Laboratories, Inc., Arthur D. Little, Inc., Biodynamics, Southwest Research Institute, U.S. Testing, Underwriters Laboratories, the University of Pittsburgh, and Weyerhauser. Weyerhauser appears to be an exception to the requirement that test laboratories have no manufacturing interest. That laboratory is prohibited from testing products of the parent company.

Applicability

The code applies to all buildings in the state with the exception of Buffalo and New York City. The buildings covered by the provisions are indicated by reference to Chapter B, all residential and nonresidential buildings, public and private including accessary structures: Chapter E, conversions, alterations additions and repairs and Chapter D, manufactured housing. This is a very broad spectrum as contrasted with many Code regulations which specify certain products under limited conditions of use. The two excepted communities have authority to establish a separate code provided it is as stringent as the state requirements. Buffalo has chosen to adopt the state-wide provisions while the city of New York has not.

Products which are covered by the provisions of the regulation include wire and cable jacket and insulation, plastic plumbing pipe, plastic conduit, duct work, insulation, and interior finish. The interior finish group is structured to cover most of the traditional items (*e.g.* ceiling materials) as well as some products which are frequently classed as furnishings (drapes). Walls, wall coverings, doors, ceiling materials, flooring, and window treatment are all considered to be interior finish for the purposes of this regulation. The impacted products are listed under paragraphs 1120.la and 1120.lb Article 15. The itemized definition of interior finish and interior floor finish as specified in sections1120.lb(5) and 1120.lb(6) of the New York Fire Prevention and Building Code are found in part 606.3. (1)

Passing Standards

Compliance with the New York State Code requires registration of products with their relevant test data. There is no passing standard or indication of acceptability. This is an unusual portion of the law; normally both the test protocol to be used and the level to be achieved by the candidate products are defined. The need to specify a passing grade is avoided by defining this as right to know legislation. The data obtained are considered to provide information to the public rather than serve as a criterion for exclusion of products from certain markets.

Registration Requirements

After being tested according to the specified version of the University of Pittsburgh test (2) products are registered in a data base which is accessible to the public. The toxicity test data, physical fire test data, halogen content and formulation are to be submitted. The formulation data are considered to be confidential and are not available to the public.

The forms required by the state are illustrated in Appendix B.

NEW YORK STATE FIRE PREVENTION CODE 9NYCRR 1120, ARTICLE 15

Registration Format

The law identifies three ways in which products may be registered. A single product may be tested and individual data submitted for registration. Alternatively, by use of a pre-approved generic category system, products may be registered without testing. If a product is composed wholly of registered materials the product may be classed as a component product. No test data are submitted for components products; instead, the data obtained from individual registered subunits are reported.

To register products as part of a generic category, a manufacturer or group of manufacturers may define groups of chemically similar products. Selected products are tested and considered to represent the group as a whole. Approval by the state is dependent on a demonstration that these categories are chemically homogeneous.

Data Base

The data are entered into a public file for reference. In most cases the "public" will be comprised of fire fighters, for whom smoke constitutes the working environment, and architects and specifiers, who have serious legal responsibility for choosing materials, and additional interested parties including competitors, scientists, regulators and inquisitive bystanders. Marketing strategies may make the data base of interest to manufacturers. The fact that the competition has not, or has not as yet, registered a product with less (or more) desirable properties with respect to toxicity does not go unnoted.

Whether the actual implementation program as it has developed provides information which is more instructive and less subject to abuse is not certain. There will certainly be problems in the conversion of raw test data to units which can facilitate realistic comparisons of products. Weight is not a relevant unit for products such as wire and cable which is measured, purchased and installed in lengths. One of the tradeoffs, which will prevent maximal use of the test data, is registration of generic groups. The products are registered separately but rely on data which were derived from other products of undefined (to the user) relationship. With a group registration a buyer cannot determine which of a series of products has the better toxicity values or which, by contrast, is least desirable by virtue of a less promising LC_{50}.

The utility of the data base to specifiers has yet to be demonstrated. No product choice would be made solely on the basis of toxicity, however, the toxicity value can be expected to augment the description of the fire performance of a product. Problems may arise due to the fact that specifiers are not accustomed to evaluating biological data in which numbers are related on a logarithmic scale. The specifier more frequently reviews information which is presented on a mathematical scale. It may be hard to accept that the relationship between 1 and 2 is the same as the relationship between 10 and 20 or 100 and 200; or that 10 is not necessarily different from 20. The potential for over interpretation of differences between products is great.

Public Access

The indication of regulatory compliance is the Department of State (DOS) registration number. This number provides linkage to the actual data file but does not itself convey any information about the product. The DOS numbers are published by the state as a book of registered products containing the manufacturers name and the product identification; again, no information about the product is included. If the user has greater interest the registered test data are available upon request. The actual documents used are available through freedom of information procedures.

NEW YORK CITY

The agency which has the primary authority in New York City is the Department of Buildings, Division of Materials and Equipment Acceptance. This unit of the Buildings Department administers a variety of codes, all of which relate to standards for product acceptability in the city. The following discussion includes only requirements pertaining to smoke toxicity.

Impacted Products

The products for which toxicity requirements are in effect in New York City include thermal insulation, walk in refrigerator panels, plastic sprinkler pipe and duct and interior finish. Regulations for carpet underlayment and floor coverings are implied as a component of interior finish and specifically covered under a separate code provision, Local Law 16.

Test and Test Laboratories

The Pittsburgh test is required by the city of New York in order to be consistent with the State's requirement. This replaces the National Bureau of Standards test, which was previously required by the city. The form of test being used is the protocol published by the state although there is some possibility that minor changes in protocol may be requested by the administrator under special circumstances. The laboratories which perform the testing must be certified by the city and for any product to be registered these laboratories are expected to submit the full test report, together with a comparison of the product performance results with the results obtained during studies of a wood standard. In the absence of the full report or of the judgement comparing the product with wood, the application will not be accepted.

The certification procedures require that the laboratory be experienced with the testing method, independent of any manufacturing sponsor, and have a qualified toxicologist to supervise testing. Laboratories which wish to become certified must demonstrate proficiency by performing tests on specimens provided by the city. The laboratories which are certified by the city of New York (May 1989) are Anderson Laboratories, Inc., Southwest Research Institute, U.S. Testing, Underwriters Laboratories and the University of Pittsburgh.

Passing Standards

For each regulated product the passing level is defined by the code as "no more toxic than wood... ." The value which is used for this comparison is the LC_{50} of the tested product with the LC_{50} of wood.

Because it has been determined that wood smoke includes a broad spectrum of products which have differing toxicities, depending apparently on the wood density, resin content and sample geometry, it has been a problem to arrive at the numerical value which will be selected to represent "wood". Under consideration has been a proposal to compare test values against a mean obtained from many species of commercially available woods. A second proposal is to choose the most toxic wood for comparison. A passing value derived from the least toxic wood would result in an almost complete ban on combustible materials in the city. The problem of identifying a passing standard has been resolved for carpets and floor coverings where there is agreement that the appropriate product for comparison is red oak flooring, the largest selling wood flooring in the city.

The unit of measure to be used for this comparison has also been the subject of discussion and agreement. For these products the LC_{50} will be expressed in units of surface area (square millimeters) rather than weight (a more common choice). The reasoning here is that both carpet and floorcovering are purchased and installed in quantities sufficient to cover a surface. For this comparison the weight of the product is not relevant. To pass the test, a carpet or flooring product must have a surface area of 1635 mm^2 or greater at the LC_{50} level. The value, 1635 mm^2, is the mean obtained for red oak flooring material as a result of testing in the Pittsburgh protocol by three of the laboratories certified by the city.

In the evaluation of test data, the laboratory must certify that the smoke from the candidate product is not more toxic to humans than the smoke from wood. This relationship is satisfied by a direct comparison of LC_{50} values when the major toxic agent can be shown to be CO (as with wood) or HCN. This is because the mechanism of toxicity of CO and HCN are both similar in the two species and because the CO/HCN toxicity ratio is the same for each. When the major toxicant works through a different mechanism and the CO/toxicant x ratio in the two species is different the direct comparison of

LC_{50} values is not satisfactory. A correction factor must be applied by the test laboratories before the comparison is made. This is the case for products which work through HCl or HF toxicity. For example, for an HCl or HF generating product it is necessary to divide the LC_{50} value by 7. The correction factor is applied because the halogen is more toxic (compared to CO) in the human than in the mouse.

Applicability

These code requirements are not as wide in application as those found in the State code. Only places of public assembly, hotel corridors, and some homes are included. Single and multiple family residences are specifically excluded.

THE PHILOSOPHY OF REGULATION

The two examples of regulatory control illustrate differing views concerning the management of a complex public safety issue. The traditional belief, as seen in New York City, is that there is a logical basis for the separation of good enough products from ones which are not adequate or are bad. The burden of specifying the acceptable limits is on the regulatory agency. The choice of that limit is made much easier when there is a natural and clean division between acceptable and unacceptable products. It is very difficult to define the boundary when the test data present a continuum from bad through good and when the absolute relationship between test data and actual fire events is difficult to authenticate. The toxicity test data present these problems to any authority trying to identify limits of acceptable performance. After this boundary is identified by the regulatory authorities, the manufacturers have only to meet the imposed limit. There is no advantage to a manufacturer in developing or marketing a product which exceeds the required standard. In fact, in many cases the test results are simply presented as passing/non-passing with the result that neither manufacturer nor user is aware of the relative merits of competing products. The manufacturer does not assume any risk in the choice of the passing standard as it is defined by an outside agency.

The aim of this traditional type of regulation is to prevent products defined as unacceptable from being used in the jurisdiction. Other than directing marketing effort to divert a failing product to a less restricted use the limit or performance test has very limited impact on products introduced into the marketplace.

By contrast, data registration realistically acknowledges that in the consideration of fire properties there is a series of product attributes which may change in importance under the conditions of product use. This format acknowledges that a regulatory body is not likely to predetermine all of the possible complex conditions surrounding the use of a product. In addition, where there is no easy demarcation between good and not good on the basis of test values the regulator may not welcome the political quagmire of defining the limits of acceptability. The registration format places the choice of product acceptability rather firmly on the manufacturer and user and moves the responsibility one step away from the regulatory agency.

The manufacturers do not have a limit of acceptability to meet; however, when the manufacturer is in contest with the skillful product liability lawyer the manufacturer finds great advantage to having marketed one of the better products of those tested and registered. In its simplest terms, the marketing and legal advantage belongs to the manufacturer having the best product. The competition for the position of "best" is one which is never over. It is always possible for another company to make an improved product and usurp the position of "best" in the marketplace. Because the state appears to consider that better products are greatly needed, there is hope that the competition for "best" product or even better than average product will find participants. The strength of marketing the best product for a particular use is one which is most apparent to manufacturers who have tangled unsuccessfully in court over just such issues.

The earliest indications from the New York State regulation are that manufacturers are finding the combined value of marketing advantage and protection from product liability to be remarkably persuasive. Three changes are apparent already:

First, old products disappeared from the market when it was shown that, in addition to other discouraging attributes, the combustion product toxicity rated at the bottom of the spectrum of tested materials.

Second, products with significantly improved fire properties, including reduced smoke toxicity, exist but may never have been marketed. These products in general are more expensive and have languished on the shelves for lack of buyer interest. A series of such products have been dusted off and made available to newly sensitized purchasers with relative success.

Third, the development of altogether new products which appear to have significant advantages over the previous generation of materials for the same use has started. In the first industries to register under these New York State codes, these new products are under vigorous development and are starting to appear.

The registration provision in the state building code provides an impetus for manufacturers to market a new generation of improved materials and gradually move the average product toward the more desirable end of the test spectrum. At the same time it provides a powerful weapon in product liability issues as a manufacturer will have no opportunity to claim ignorance of the performance of registered products.

REFERENCES

1. *New York State Uniform Fire Prevention and Building Code 9NYCRR Article 15, Part 1120, Combustion Toxicity Testing*. Albany, New York, December 1986.
2. *New York State Uniform Fire Prevention and Building Code, Regulations for Implementing Building Materials and Finishes Fire Gas Toxicity Data File* (Statutory Authority: Executive Law 91) 9NYCRR 439.
3. Anderson R. C., K. M. Daring, M. Long. *Study to Assess the Feasibility of Incorporating Combustion Toxicity Requirements into Building Material and Furnishing Codes of New York State. Volume 3, Final Report to Department of State Department of Fire Prevention and Control.* Arthur D. Little, Inc. May 1983.
4. National Bureau of Standards Report NBSIR 82-2532, June 1982. *Further Development of a Test Method for the Assessment of the Acute Inhalation Toxicity of Combustion Products.*

APPENDIX A

ARTICLE I5
PART 1120 COMBUSTION TOXICITY TESTING
9 NYCRR 1120
Effective Date: December 16, 1986

§1120.1a General Requirements. No covered product shall be installed or used in or as part of a building which is subject to Chapter B, Chapter E, or factory manufactured homes under Chapter D of the Code unless (1) such covered product has been tested in accordance with the toxicity test method prescribed in Section 1120.2 and (2) a report of such test has been filed with the Department of State in accordance with Section 1120.3. The covered products regulated by this Part are as follows: beginning one year after the effective date of this Part, any electrical conduit or electrical wire insulation; beginning two years after the effective date of this Part, any pipe, duct or thermal insulation; and beginning three years after the effective date of this Part, any interior finish or interior floor finish.

§1120.1b Definitions. For the purpose of this Part:

§1120.1b-1 "covered product" means a product regulated by this Part which is distinguished from other covered products by properties other than color, shape or dimensions.

§1120.1-2 "electrical conduit" means any pipe or tube which is used to carry electrical wiring of any voltage or amperage and consists of synthetic or natural polymers not including pipes and tubes within appliances, fixtures or mechanical equipment sold as assemblies.

§1120.1b-3 "electrical wire insulation" means any covering, coating or wrapping material which (i) is used to provide a heat barrier and or an electrical conductivity barrier; (ii) is applied to electrical wiring designed to carry electrical current of any voltage or amperage; and (iii) is *not* within an appliance, a fixture or mechanical equipment sold as an assembly.

§1120.1b-4 "duct" means any duct work which is located within a building.

§1120.1b-5 "interior finish" means any product defined as such in Section 606.3 of the Code.

§1120.1b-6 "interior floor finish" means any product defined as such in Section 606.3 of the Code.

§1120.b.-7 "person" means an individual, partnership, corporation or association.

§1120.1b-8 "pipe" means any pipe which is located within a building, not including pipes which are part of appliances, fixtures or mechanical equipment sold as assemblies.

§1120.1b-9 "product" means any material or combination of materials consisting of synthetic or natural polymers, commercially available in end use form for installation in or as part of a building or used in a building as electrical conduit, electrical wire insulation, ductwork, interior finish, interior floor finish, thermal insulation or a pipe.

§1120.1b-10 "thermal insulation" means any material which is used principally as a thermal barrier to heat transmission, not including materials used in appliances, fixtures or mechanical equipment sold as assemblies.

§1120.2a Test Protocol. The combustion toxicity test required by Section 1120.1a shall be conducted in accordance with the "Procedure for Evaluation of Acute Toxicity Resulting from Exposure of Mice to Thermal Decomposition Products Using a Modification of the Protocol Developed at the University of Pittsburgh" as set forth in Section 1120.7.

APPENDIX A (continued)

ARTICLE 15
PART 1120 COMBUSTION TOXICITY TESTING
9 NYCRR 1120
Effective Date: December 16, 1986

§1120.2b Testing Organizations. Any test required by Section 1120.1a shall be conducted by a laboratory acceptable to the Secretary of State. The list of acceptable laboratories shall be made available to any person upon request.

§1120.3 Test Reports. (a) The report required by Section 1120.1a shall be submitted to and be in a form acceptable to the Secretary of State and accompanied by an affirmation by a laboratory accepted pursuant to Section 1120.2b that the tests have been conducted in full compliance with the test protocol prescribed in Section 1120.2a. (b) The report shall contain the information specified in Section 1120.7g–4. (c) where any material required to be tested pursuant to Section 1120.1a is also required by this Code to be tested for ignitibility or flame spread, the result of such tests shall be included in the test report which is filed with the Department of State. (d) A producer or group of producers may propose to the Secretary of State to treat certain similar or related covered products as a class. Any producer or group of producers proposing such a classification must submit documents supporting and justifying the proposal and any additional information requested by the Secretary of State. The Secretary will review such submissions and approve those classifications that are deemed to be consistent with the purposes of this Part. Where a class has been approved each producer must submit, on a form prescribed by the Secretary of State, a list of all its products that belong to the class and a copy of the test report for the class. The filing by a producer of the test report as required by this section shall satisfy the requirements for all of its products in the class.

§1120.4 Data File. (a) The Department of State shall compile all test reports filed in accordance with this Part into a Building Materials and Finishes Data File, which shall be a public record available to any interested person. (b) *Release of Data Reports.* All test results released from the data file shall contain the following statement:

This test is a measure of acute toxicity of the thermal decomposition by-products of the tested material using a specified procedure under controlled laboratory conditions. The test results alone do not constitute a characterization of the hazard, safety or risk of materials, products or assemblies under actual fire conditions. The results of this test, if used in any assessment of hazard or risk, should be considered in conjunction with all of the factors which are pertinent to an evaluation of the fire hazard of a particular end use.

§1120.5 Certification. No person shall be liable for a violation of this Part if such person obtains a certification in such form as the Secretary of State by rules or regulations may prescribe that the product has been tested pursuant to the method prescribed in Section 1120.2 and that the report of such test has been filed with the Department of State pursuant to Section 1120.3.

§1120.6 Violations. (a) It shall be a violation of this Part for any person to make, with respect to any product, a false certification. (b) No person who in good faith relies upon a certification of a manufacturer pursuant to Section 1120.5 of this Part shall be deemed in violation of this Part

§1120.7 Procedure for Evaluation of Acute Toxicity Resulting from Exposure of Mice to Thermal Decomposition Products Using a Modification of the Protocol Developed at the University of Pittsburgh.

§1120.7a Purpose of Procedure. To obtain required data on lethal potential of the thermal decomposition products of the test material.

APPENDIX A (continued)

§1120.7b Test Director. The Director of each test shall be identified.

§1120.7c Test Article. The test article shall be as completely identified as is possible. Identifying information shall be obtained from the supplier of the test article. A physical description (including, but not limited to, dimensions, color and weight) shall be entered into the record.

§1120.7d Animals.

§1120.7d-1 Species. The species of animal used shall be the Swiss Webster mouse.

§1120.7d-2 Source. Animals shall be obtained from any animal breeding laboratory.

§1120.7d-3 Sex. Male mice shall be used.

§1120.7d-4 Size. Mice shall weigh 25-28 grams at time of exposure. (Smaller mice may be able to crawl into the exposure chamber; larger ones may not be able to breathe normally in the apparatus.)

§1120.7d-5 Housing. (i) Animals shall be quarantined for a minimum of one week prior to use.
(ii) Animals shall be group-caged in regular plastic cages with bedding.
(iii) The cage shall be thoroughly cleaned and sanitized prior to use and provided with any commercially available bedding.
(iv) Food and tap water shall be freely available. Any commercially available rodent chow is acceptable. Neither food nor water need be analyzed for specific content.

§1120.7e Test Equipment.

§1120.7e-1 Animal Exposure Chamber. To accommodate 4 mice. This is a locally manufactured item. It is made of glass and can be produced by any competent glass blower, starting with two (one male and one female) 103/60 joints. These are available from any glass distributing firm. With these as starting pieces, the glass blower can easily manufacture the animal exposure chamber. The specific dimensional requirements of the chamber are set forth in Attachment B.

§1120.7e-2 Crucible Furnace and Furnace Programmer.
(i) The furnace and programmer system must be capable of heating at 20°C/minute with linearity to 1000°C. Any system substantially meeting the specifications of the modified furnace and programmer outlined in (ii) and (iii) is acceptable. The use of product names in this Part is for descriptive purposes only and is not to be construed as an endorsement of that product.
(ii) The University of Pittsburgh uses Lindbergh Box Furnace Model 51894-S. This furnace has operating capacity up to 1100°C. Chamber dimensions: 229mm x 229mm x 357mm. Rated at 3500 watts on 240 volts, 60hz. A 21mm hole with a ceramic sleeve is put in the door of the furnace. A 13mm hole is put in the bottom to accommodate the pedestal and a 13mm hole is put in the rear wall to place a thermocouple for the recorder above the Platinel II thermocouple for the programmer. (These modifications can be made by the furnace manufacturer.)
(iii) The University of Pittsburgh uses Stanton Redcroft Variable Rate Linear Temperature Programmer Model Mark II LVP. Temperature range to 1100°C, rate of climb 2-40°C/minute. This programmer incorporates three term controllers which have their power outputs continuously variable via solid state thyristors. The instrument has a temperature deviation meter to provide a visual reference of the chamber temperature/set point temperature relationship. Maximum current to furnace is 40 amps. Temperature control system is for use on 240 volts, 60hz.

§1120.7e-3 Sample Holder. This is the container in which the material is placed for thermal decomposition within the furnace.

APPENDIX A (continued)

This is a standard laboratory item available through any scientific catalog.

§1120.7e-4 **Pedestal.** The pedestal is constructed of a quartz tube and pedestal. The construction extends from the weight sensor to the inside of the furnace. Specifications for this construction are illustrated in Attachment C. The lower end is stabilized by a glass support attached to the weight sensor.

§1120.7e-5 **Flowmeters.** Two flowmeters are necessary: one meter to have the capacity to adequately measure at least 9 liters per minute; and one meter to have the capacity to adequately measure at least 20 liters per minute. These are standard lab items available through any scientific catalog.

§1120.7e-6 **Pump.** The pump should have the capacity to draw at least 20 liters per minute through the exposure chamber. This is a standard lab item available through any scientific catalog.

§1120.7e-7 **Balance for Sample and for Animals.** This is a standard catalog item. Any top loading balance or any other scientifically capable weight sensor is acceptable.

§1120.7e-8 **Weight Sensor.** Placed beneath the furnace, this is the scale upon which the pedestal rests. It is connected to the recorder to record the weight of the sample material as it is thermally decomposing. A glass support is attached to the center of the pan to accommodate the pedestal.

§1120.7e-9 **Calibration Weights.** These are standard catalog items.

§1120.7e-10 **Filter System.** This is an activated charcoal filter which is laboratory constructed to clean the exhaust from the test chamber before it is discharged into the atmosphere.

§1120.7e-11 **Thermometer or Thermocouple.** To indicate the temperature within the exposure chamber, any ordinary laboratory thermometer can be used.

§1120.7e-12 **Recorders.** These are standard strip chart recorders available through a scientific catalog. One is used to provide a continuous recording of the temperature in the chamber and of the weight of the sample. A separate thermocouple is needed for this recorder. Another records carbon dioxide and oxygen. A third records carbon monoxide output. Many manufacturers produce recorders that would be acceptable.

§1120.7e-13 **Meters.** Three meters are required: 1 O_2 meter capable of measuring 0–22 percent oxygen; 1 CO_2 meter capable of measuring 0–5 percent carbon dioxide; 1 CO meter capable of measuring 0–5 percent carbon monoxide. These are available through standard scientific catalogs.

§1120.e-14 **Quartz Tube.** This tube extends from the furnace and is 46 cm in length. This quartz tube is available from several glass blower suppliers. A male ball joint should be provided on the end of the quartz tube which extends out from the furnace. Such fitting can be attached by any competent glass blower. Dimensional details of this quartz tube may be found in Attachment C.

§1120.7e-15 **Pyrex Connector.** This connector extends from the quartz tube to the exposure chamber. This is a stock item for a glass blower, who can modify it for these purposes. Such connectors should be fitted on one end with a female ball joint and on the other end with an O ring joint. A side opening should be provided to allow for the introduction of cool air. A detailed description of this Pyrex connector is found in Attachment C.

§1120.7e-16 **Copper Cooling Coil.** This is a laboratory item made of standard copper tubing which extends through an ice bath. The purpose of which is to introduce cool air through the Pyrex connector into the exposure chamber.

NYS DEPARTMENT OF STATE
Fire Gas Toxicity Data File
P.O. Box 115
Albany, NY 12260-0115

DOS USE ONLY	FILE NUMBER

Fire Gas Toxicity

LEFT MARGIN tab tab tab tab tab tab tab

PART A. MANUFACTURER'S PRODUCT REPORT (TYPE IN ALL CAPS)

NAME OF MANUFACTURER 1/			
DIVISION NAME OR U.S. DISTRIBUTOR, IF FOREIGN PRODUCER 2/			
MANUFACTURER'S ADDRESS (NUMBER AND STREET) 3/	(CITY)	(STATE)	(ZIP)
MANUFACTURER'S TOXICITY TEST COORDINATOR 4/	TELEPHONE NUMBER (AREA CODE AND EXTENSION) 5/ ( )		
CSI NUMBER 6/	NYS DOS CLASS APPROVAL NUMBER 7/	LAB TEST NUMBER 8/	

COMPONENT DOS FILE NUMBERS (SEPARATE WITH SEMICOLONS ";")

9/ ....

NAMES UNDER WHICH PRODUCT WILL BE MARKETED OR USED IN NEW YORK STATE (SEPARATE WITH SEMICOLONS ";")

10/ ....

MANUFACTURER'S PRODUCT IDENTIFICATION NUMBERS (SEPARATE WITH SEMICOLONS ";")

11/ ....

HALOGEN CONTENT

12/ % FLUORINE 13/ % CHLORINE 14/ % BROMINE 15/ % IODINE

NUMERICAL FLAME SPREAD RATING	SOURCE	CRITICAL RADIANT FLUX	SOURCE
16/		17/	

OTHER INFORMATION RELATED TO DESIGNATION OF PRODUCT PROPERTIES (LIMIT, 250 CHARACTERS)

18/ .... END PICA (10 PITCH) END ELITE (12 PITCH)

NOTE: Manufacturer must complete Part C and sign the certification.

664309-001 (5/87)

NCS E-25578:321

NYS DEPARTMENT OF STATE
Fire Gas Toxicity Data File
P.O. Box 115
Albany, NY 12260-0115

DOS USE ONLY	FILE NUMBER

Fire Gas Toxicity

LEFT MARGIN tab tab tab tab

PART B. PRODUCT TEST REPORT LABORATORY TEST DATA (TYPE IN ALL CAPS)

NUMBER OF SAMPLES TESTED	LC_{50}	95% CONFIDENCE INTERVAL	FURNACE TEMPERATURE AT 1% SAMPLE MASS LOSS
1/	■■■■■ 2/ g	■■■■■■■■■■■ 3/ g	■■■■■■■■ 4/ °C

FURNACE TEMPERATURE RANGE OF MOST RAPID MASS LOSS	MEAN FURNACE TEMPERATURE AT SPONTANEOUS FLAME	RESIDUE (SAMPLE AVERAGE)
■■■■■■■■■■■■■■■■■■■■ 5/ °C	■■■■■■■■■■■■■■■■ 6/ °C	■■■■■■■■ 7/ %

FOR ITEMS 8-16, FROM A SINGLE RUN AT OR NEAR THE LC_{50} TEST SAMPLE MASS:

8/ ppm	Maximal concentration of carbon monoxide in the exposure chamber
9/ °C	Furnace temperature at the point of maximal carbon monoxide
10/ %	Maximal concentration of carbon dioxide in the exposure chamber
11/ °C	Furnace temperature at the point of maximal carbon dioxide
12/ %	Minimal concentration of oxygen in the exposure chamber
13/ °C	Furnace temperature at the point of minimal oxygen
14/	Number of times the exposure chamber temperature exceeded 45°C
15/ sec	Average duration of exposure chamber temperatures in excess of 45°C
16/	Eye condition of test animals (select one): (1)All apparently normal (2) Some apparent damage (3) Some severe damage

PHYSICAL DESCRIPTION OF THE TEST LC_{50} SAMPLE USED

17/ ....

NOTES AND OBSERVATIONS

18/ ....

LAB TEST NUMBER	NAME OF TESTING ORGANIZATION
19/	■■■■■■■■■■■■■ 20/

TEST DIRECTOR OR COORDINATOR	TELEPHONE NUMBER
21/	22/ ( )

LABORATORY AFFIRMATION — I affirm that the test, the results of which are reported herein, was conducted in compliance with 9 NYCRR 1120.2(a).

Signature X ______________________ *Date* ______________

Name ______________________ *Title* ______________________
(PRINT OR TYPE)

NYS DEPARTMENT OF STATE
Fire Gas Toxicity Data File
P.O. Box 115
Albany, NY 12260-0115

DOS USE ONLY FILE NUMBER

Fire Gas Toxicity

LEFT MARGIN tab tab tab

PART C. MANUFACTURER'S PRODUCT ANALYSIS

Represent product recipe/chemical composition (weight percentage):

1/ ....

☐ "X" BOX AND COMPLETE ITEM 8 ON REVERSE, IF ANY PORTION OF THIS PRODUCT REPRESENTATION IS A TRADE SECRET

☐ "X" BOX AND COMPLETE ITEM 8 ON REVERSE, IF DISCLOSURE OF HALOGEN CONTENT WOULD COMPROMISE A TRADE SECRET

2/ % FLUORINE 3/ % CHLORINE 4/ % BROMINE 5/ % IODINE

PRODUCT DESCRIPTION

6/ ....

EXPECTED USES FOR WHICH PRODUCT IS MANUFACTURED

7/ ....

☐ "X" BOX, IF PRODUCT TEST INFORMATION IS FILED ELECTRONICALLY; ELECTRONIC I.D.: ______

MANUFACTURER CERTIFICATION — I certify, subject to penalty for its falsity, that the information furnished herein, and in any electronic filing identified above, is true and accurate to the best of my knowledge.

Authorized Signature X ______ *Date* ______

Name ______ (PRINT OR TYPE) *Title* ______

NYS DEPARTMENT OF STATE
Fire Gas Toxicity Data File
P.O. Box 115
Albany, NY 12260-0115

DOS USE ONLY	FILE NUMBER

Fire Gas Toxicity

LEFT MARGIN

PART C. MANUFACTURER'S PRODUCT ANALYSIS (continued)

Specify information or portions thereof for which exception from disclosure is requested, and state the reasons why such information should be excepted.

8/

☐ "X" BOX, IF THE DISCLOSURE OF ANY PORTION OF THE RESPONSE TO THIS QUESTION WOULD DEFEAT THE PURPOSE FOR WHICH THE EXCEPTION IS SOUGHT

Chapter 7

A REVIEW OF THE COMBUSTION TOXICITY OF POLYMERS BY CLASS

Shayne C. Gad

This chapter is written with the aim of providing an overview of the available information as to the combustion toxicity of the major classes of polymers. Such a review will always become dated rather quickly and will always miss at least some minor sources. However, it is impressive to see how much data is available, and at least a snapshot can be effectively captured.

Given all the groups of people involved in the area of combustion toxicology, it should not be a surprise that there is also disagreement over how the results of the different test systems are evaluated. Or, more to the point, how they should be used as a guide to where society goes from here.

Most of the differing opinions come from the various research groups involved. However, there are at least two other sets of opinions. The first of these, expressed most recently by J. O. Punderson (Punderson, 1981), basically states that because none of the test systems proposed to date are perfect and because a program of full scale fire validation studies has not been performed, we should not at this time utilize the results from laboratory scale tests in making decisions about the safety of materials in fires. A secondary argument that is raised is that it would be wrong to assume that a material's potential to evolve toxic gases in a fire situation is directly equivalent to its hazard in a fire situation. Both of these points are valid, in so far as they go, but not to the extent of the authors' conclusion. The test systems available are not, as pointed out earlier, perfect. But the major systems do, as we shall see shortly, produce data which is relatively reproducible and consistent from one system to another. No program of large scale tests has been performed to validate the laboratory scale results but supporting data from some real fires and from some large scale tests does exist, and does not contradict the results of laboratory scale studies (rather, it appears to support them). The third point is one which the authors have raised earlier - decisions cannot be based on combustion toxicity alone. Rather, conditions of use and other hazard potentials (flamability and combustibility, for example) must be included in these decisions. Knowing this should not, however, prevent us from making use of the combustion toxicity data we have, together with these other pieces of information. It also does not mean that in some cases the potential for the production of toxic gases may not be so great that this may be the most important factor in decision making.

The alternative viewpoint, from a nonresearcher's viewpoint, is that presented by regulatory groups such as the MEAD (Material Equipment Acceptance Division) of the City of New York, which back in the late 1970's certified several laboratories around the country to perform combustion toxicity evaluations for them. These evaluations were on materials to be used in buildings in the City of New York, and materials were passed or failed (accepted or rejected) for such use based purely on whether their LC_{50} (in each of three modes) were greater or less than those of Douglas Fir. If any of these six figures were less than those of wood (remembering that the lower an LC_{50} or IC_{50}, the more toxic it is), then the material was rejected for use. This attitude is as inappropriate as the first; the data is not that precise (22.0 mg/L is not significantly different from 23.0 mg/L) and other aspects of a hazard evaluation are not included, though they are very relevant.

Among the active researchers in the field, there are three proposed systems for evaluating (and making "regulatory" decisions based on, noting that in this context "regulatory" may refer to decisions actually made internally by a company) combustion toxicity data. The National Bureau of Standards (now NIST) has proposed materials be classified as:

a. not significantly different from wood, with all LC_{50}s greater than 1/2 or 1/3 of that of wood
b. more toxic than wood, with no LC_{50} less than one-tenth that of wood, but at least one LC_{50} less than one-half or one-third that of wood and
a. extremely toxic, with one or more LC_{50} values less than one-tenth that of wood.

The second system proposed by a group of researchers is that proposed by the University of Pittsburgh group (Alarie and Anderson, 1981) which proposes the use of a set of curves (as shown in Figure 7-1) which interrelate the LC_{50} (on the horizontal axis) and the LT_{50} (time expected to kill 50% of exposed animals at that concentration, on the vertical scale) of materials to classify them as from Class A (not significantly different from wood) to Class D ("supertoxic"). This approach has the very real advantage of a list incorporating the concept of speed of effect, but also has the drawback that it does not necessarily address incapacitation at all.

Lethality-Response (LC50)

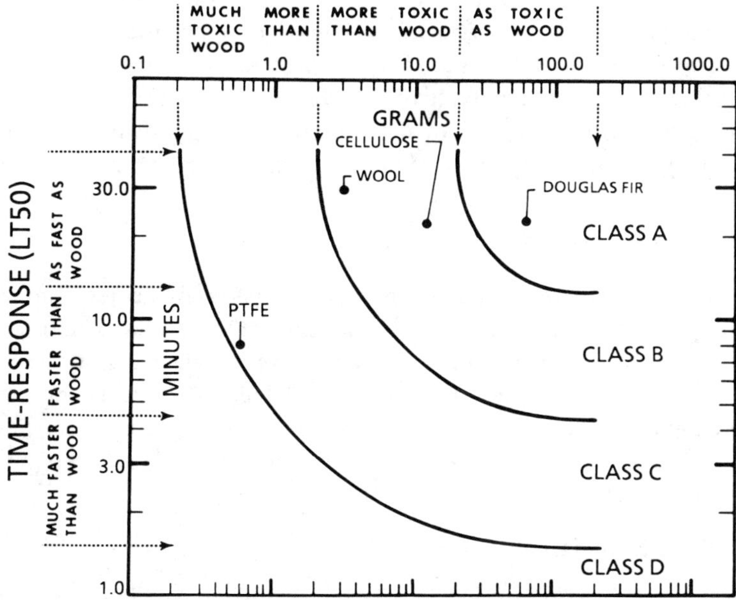

Figure 7.1

Finally, the CMIR group proposed a system which evaluated the data based on statistically significant differences in LC_{50}, IC_{50} and IT_{50} (time to incapacitation for 50% of the exposed population at the IC_{50} concentration). This system has the dual advantages of utilizing all the available data on a material (the worse case from each of the three modes - flaming, nonflaming, and 440°C - where data is collected) and of "rewarding" the more accurate and reproducible test methods by allowing them to establish finer differences.[1] Because of the nature of the data and the desire to make conservative evaluations, the scheme presented in Figure 7-1 is used to sort each material into its

1 This "rewarding" is based on the fact that the smaller the variation in repeated measures, the less real difference is necessary for statistical analysis to serve to establish a difference.

relative position compared to all other materials evaluated, with actual statistical testing being performed based on the LC/IC and IT_{50}s and their 95% confidence intervals. First comparisons are by simple Student t-tests, but the final comparisons are by a series of Wilcoxon rank sum tests, or a Kruskal-Wallis nonparametric analysis of variance, as appropriate. This is the methodology used, so far as possible, in those actual comparisons of data which are performed in later parts of this chapter.

This scheme has the additional advantage of using all the information available, without attempting to impart greater accuracy than is actually present.

RESULTS TO DATE (BY CLASS OF POLYMER)

Having briefly reviewed the methodologies and the techniques for evaluating the resulting data, we can now review the existing data base. The quality of information available is uneven. However, viewed as a composite, it should serve to enable one to make some initial judgments and comparisons. To the extent possible, the data presented here are grouped by class. Several reviews of individual classes exist in the literature, and are referred to under the appropriate class.

REFERENCE MATERIALS

Since we need a set of benchmarks against which to compare the different classes, it is necessary to first examine the available reference materials. Much of the concern regarding the greater hazard potentials of plastics in a fire is based on the concept that traditional building materials generate less toxic thermal gradation products - in fact, that such products have their toxicity defined (or explained) by oxygen deprivation and carbon monoxide production. As a consequence, almost all the reference materials (or standards) proposed are "traditional" materials used in construction and furniture - various types of wood, compressed wood, cotton or wool. For all these except wool, oxygen deprivation and carbon monoxide production do, in fact, explain the combustion toxicity rather well. Wool, however, also produces a considerable amount of hydrogen cyanide - a fact which, as will be shown below, is well reflected in the toxicity of its thermolysis products.

The most widely recognized of the standards are natural woods (most prominently Douglas fir). The literature (O'Mara, 1974; Einhorn, 1977; Nelson *et al.*, 1978) indicates that a number of compounds (other than CO and CO_2) are present in the thermolysis product streams from wood, though only acrolein may be of toxicological concern. Though a wide variety of natural woods has been examined, no substantial differences have been found be-

Table 7-1. Combustion Toxicity Of Natural Woods

Wood Type	Evaluation System/Mode		Toxicity	Reference
Douglas Fir	University of Pittsburgh		LC = 63.8 "gram load" LT = 22 minutes	Alaire and Anderson, 1979
Compressed Wood	University of Pittsburgh		LC = 48.7 "gram load" LT = 19 minutes 1979	Alaire and Anderson, 1979
Plywood	NBS	Pyrolysis Flaming	LC_{50} = 26.9 mg/L LC_{50} > 26.9 mg/L	Potts and Lederer, 1977
Red oak and Yellow Pine	NBS	Pyrolysis	LC_{50} = approximately the same as plywood	Potts and Lederer, 1977
Wood (type undefined)	DIN	Pyrolysis	LC_{50} = 29 mg/L	Kimmerle and Prager, 1980
Douglas fir	NBS	Nonflaming Flaming	LC_{50} = 23 mg/L IC_{50} = 18 mg/L LC_{50} = 26 mg/L IC_{50} = 14 mg/L	Gad and Smith, 1983
Douglas fir	CMIR	Nonflaming Flaming	LC_{50} = 24.3 mg/L IC_{50} = 17.2 mg/L LC_{50} = 26.7 mg/ LC_{50} = 13.8 mg/L	Gad and Smith, 1983
Hem fir	CMIR	Nonflaming Flaming	LC_{50} = 26.3 mg/L IC_{50} = 12.6 mg/L LC_{50} = 27.2 mg/L IC_{50} = 9.4 mg/L	Gad and Smith, 1983

Note: The University of Pittsburgh system, because of its "dynamic" (flow through) nature does not yield LC_{50} or LT_{50} results, but rather lethal or incapacitating concentration or times.

tween the different types. Table 7-1 summarizes the existing toxicity data for natural woods.

An important point which emerges, not just form studies of these standards, but also of synthetic polymers, is that the rate of heating in a test system (and, assumedly in any real fire) has a marked influence on both the nature of the decomposition products and their toxicological impact. As a general rule, the faster the rate of heating the greater the "toxicity" of the evolved combustion products. This relationship is not such, however, as to be explainable as just a decomposition rate phenomenon (Gad and Smith, 1983).

Table 7-2. Combustion toxicity of proposed reference standards other than wood

Reference Material	Evaluation System	Toxicity	Reference
Surgical cotton	University of San Francisco	1 gram load T_D 13.90 (±3.51) T_I 9.37 (±1.21)	Hilado and Solis, 1977
Fiberglass Insulation	University of Pittsburgh	LC = 35.7 gram load LT = 25 minutes	Alaire and Anderson, 1979
wool (undyed)	University of Pittsburgh	LC = 3.0 gram load LT = 27 minutes	Alaire and Anderson, 1979
Wool	NBS	Nonflaming: LC_{50} = 18 mg/L LC_{50} = 16 mg/L Flaming: LC_{50} = 19 mg/L IC_{50} = 17 mg/L	Previously unpublished data, S. C. Gad
Wool	DIN	LC_{50} = 8 mg/L	Kimmerle and Prager, 1980
	DIN	LC_{50} = 24 mg/L	Kimmerle and Prager, 1980
Cotton	University of Pittsburgh	LC_{50} = 28.3 mg/L	Alarie *et al.*, 1983
Wool	CMIR	Nonflaming: LC_{50} = 9.4 mg/L LC_{50} = 7.6 mg/L Flaming: LC_{50} = 8.3 mg/L IC_{50} = 7.8 mg/L	Previously unpublished data, S. C. Gad
Wool	NBS	Nonflaming: 15.8 - 45.1 mg/L Flaming: 23.8 - 58.3 mg/L	Levin *et al.*, 1982

Note: The University of Pittsburgh and University of San Francisco systems, because of their "dynamic" (flow through) natures do not yield LC_{50} or LT_{50} results, but rather lethal or incapacitating concentration or times.

Data concerning the other materials, which have been proposed as reference standards, are presented in Table 7-2. In addition to the traditional building materials (cotton and wool), several workers (O'Mara, 1974) have proposed using one of several synthetic materials (α-cellulose and fiberglass), because these are better defined and constant in their initial composition.

ABS (ACRYLONITRILE - BUTADIENE - STYRENE)

Notwithstanding an extensive literature, and concern in terms of potential for carcinogenicity of the component monomers from this class, there is only a limited body of combustion toxicity data. Kourtides *et al.*, (1978), in their own systems, reported an LC_{50} (for 30 minutes of exposure to a pyrolysis product) of 10.45 (± 0.02) mg/L. Alaire and Anderson (1981) reported an LC_{50} in their system (where wood is 63.8) of 6.3 grams with an LT_{50} (time to lethality) of 9 minutes (as opposed to 22 minutes for wood). A second ABS in the same system yielded an LC_{50} of 8.32 (Sangha *et al.*, 1981). Kimmerle and Prager (1980), however, reported an LC_{50} of 48 g/m^3 (as opposed to, in their system, 29 for wood), with a flame retardant and 29 without a flame retardant.

Zitting and Savolainea (1980) have reported that, at both 300 and 350°C, the three most prominent components of ABS decomposition streams were, in order of importance, styrene, aldehydes, and carbon monoxide. Exposure to such streams was correlated with decreased O-deethylation activity in the liver and kidney.

In the NBS system, Levin *et al.*, (1982) reported LC_{50} values of 15.6-22.1 mg/L (flaming) and 22.0 to 38.0 (nonflaming).

In summary, there are limited data available on ABS decomposition. An overall evaluation based on these limited data indicates that though the decomposition products of ABS are more of a hazard than wood, the difference is not of great significant.

ACRYLICS

There is now more information available for this class than for many others, with Johnston *et al.*, (1988) having published a separate review. In general, they found modacrylic combustion products to be more acutely toxic than those of other acrylics or of wood. Table 7-3 summarizes much of the available data from the five major test systems.

Alaire and Anderson (1981) also reported (for modacrylic), an LC_{50} of 4.9 mg/L and an LT_{50} of 18 minutes. Kimmerle and Prager (1980) have reported separate LC_{50} values for an acrylic and a modacrylic - but both values were 6 g/m^3 (which equates to 6 mg/L). Finally, Zaborowski *et al.*, (1972), reported that orlan fabric (an acrylic) generates HCN upon thermal decomposition.

In summary, all the data for this class (though it is not an extensive set), indicates that their thermolized products are at least four times as toxic as wood and are more acutely toxic than those of the ABS group. As with most other polymers, as decomposition temperatures (and therefore rates) are in-

Table 7-3. Acute Toxicity of Combustion products of acrylic materials in five test systems

Test method	Conditions and exposure	Material tested	LC_{50} (mg/L)		References
DIN	Temperature, 500°C. Airflow, 200 L/min.	45% modacrylic, 55% acrylic carpet	27.0 ± 2.0		Herpol and Vandervelle, 1983
NBS	Flaming and Nonflaming	Modacrylic	7.1 (flaming)		Levin *et al.*, 1980
			4.0 - 5.1 (flaming) 4.5 - 6.0 (nonflaming)		Birky *et al.*, 1980
		Polyacrylonitrile	38.7		Williams and Clark 1983
University of Pittsburgh	Temperature, 25-600°C. Airflow, 20 L/min.	Modacrylic	8.2		Alarie and Anderson, 1981
			t_i (min.)	t_d (min.)	
FAA	Temperature, 600°C. Air recirculation, 4 L/min.	Fire retarded PMMA	7.5	14.6	Crane *et al.*, 1977
		PMMA	**	**	
		modacrylic	1.10	2.24	
USF/ NASA	Temperature, 800°C.	PMMA	3.71 ± 0.51	7.74 ± 1.79	Hilado *et al.*, 1977a
		Acrylonitrile rubber	2.19 ± 0.13	4.34 ± 0.18	Hilado *et al.*, 1977b
		Poly-methacrylimide	4.92 ± 0.06	8.46 ± 0.16	Hilado and Huttinger, 1981
	Temperature, 200-800°C.	PMMA	13.14 ± 0.38	15.58 ± 0.23	Hilado *et al.*, 1977a
			8.43 ± 0.99 9.94 ± 0.83	18.60 ± 1.91 16.73 ± 1.28	Hilado and Huttinger, 1981
		Acrylonitrile rubber	9.86 ± 0.18	15.85 ± 0.97	Hilado *et al.*, 1977b
		Poly-methacrylimide	5.94 ± 3.00	11.45 ± 2.31	Hilado and Huttinger, 1981

* Levin *et al.* (1982) reported nonflaming values ranging from 7.8 to 23.8 mg/l, with the value going up as temperature of decomposition went down.

** No incapacitation or deaths reported

t_i Time to incapacitation.

t_d Time to death

creased the "toxicity" of the products increases (Saito, 1978) though this probably actually reflects time course/concentration effect.

ALKYDS (POLYESTERS)

Yoshida *et al.*, (1978) has reported work on one material in this class, which he found to have a one-hour apparent lethal concentration 50 (ALC_{50}) in mice of 30.5 mg/L. The decomposition products were almost entirely carbon monoxide and acetaldehyde, with the carbon monoxide predominating in a 3:1 ratio. Blood gases from dead animals confirmed a level of COHb which would be expected to be lethal.

Alaire and Anderson (1981) reported data on two unsaturated polyester resins, identified as I and II. The LC_{50} values for the two were 34.8 and 57.4, respectively, while the LT_{50} values were 14 and 18 minutes respectively. A styrenated and halogen-modified form tested at the same time, however, showed an LC_{50} value of 14.4 and an LT_{50} value of 16 minutes. In 1983, Alaire *et al.*, reported an LC_{50} of 28.0 for a separate sample of polyester.

Miller *et al.*, (1976) reported some interesting work on polyesters, in particular how alterations in their structures corresponded to alterations in both flamability and smoke generation. Propylene glycol-based resins were found to generate the most smoke and were found to be most difficult to retard flamability in. The acid component used in the process also affects smoke generation.

Overall, though the data are very limited, it seems clear that the alkyds themselves generated combustion products which are only minimally more toxic than those of wood.

CELLULOSICS

This is the first of several classes of materials (other than reference materials) in this review which were found to have a significant early body of work but which does not seem to have been of particular interest during the last ten years.

Robles (1968) has reported that the primary decomposition products of cellophane were, in descending order: CO_2, NO_2, NO, COS_2, SO_2, methanol, formaldehyde, acrolein, and then a number of unidentified hydrocarbons and esters. The toxicity evaluations of these products are summarized in Table 7-4.

In summary, the cellulosics seem to produce combustion products which are moderately more toxic than those produced by wood, but less so than the ABS group.

Table 7-4. Combustion toxicity of cellulosic plastics			
Material	Reference	Conditions/system	Reported results
Cellophane	Kourtides *et al.*, 1979	Pyrolysis	LC_{50} = 10.75 ± 2.25 mg/L
Cellulose fiber insulation	Alaire and Anderson, 1979	University of Pittsburgh system	LC_{50} = 11.9 mg/L LT_{50} = 21 minutes
Shredded cellulose	Previously unreported data, S. C. Gad	NBS system	Flaming: LC_{50} = 18.4 mg/L Nonflaming: LC_{50} = 14.9 mg/L
Shredded cellulose insulation (with alum added)	Previously unreported data, S. C. Gad	NBS system	Flaming: LC_{50} = 27.8 mg/L Nonflaming: LC_{50} = 23.4 mg/L

FLUOROPOLYMERS

Extensive work has been performed on the materials in this class - probably the most extensive package of work on any polymer class, though much of this has been concentrated on one member of the group.

PTFE (polytetrafluoroethylene) is probably the single most studied material in the field of combustion toxicology. Its exact decomposition products are believed to be markedly dependent on temperature and on the oxygen available in the decomposition atmosphere. Paciorek *et al.*, (1973) reported that, in a pure oxygen atmosphere, only COF_2 and CF_4 were formed. In "room air", however, C_2F_4 and a multitude of additional forms were also generated. At temperatures above 650°C disproportionation of more complex molecules to CO_2 and CF_2 appeared to take place.

Hilado and Schneider (1979) reported that the components of the PTFE decomposition stream were dependent on the construction material used in the tube furnace used in their studies.

Arito and Soda (1977) compared PTFE and polyfluorethyl propylene (PFEP) as to the chemical nature of their decomposition products. In low oxygen conditions, both materials produced octofluorisobutylene, tetrafluorethylene, hexafluoropropylene, and octofluorisobutylene. In an air stream (*i.e.*, with more oxygen available), the main products became carbonyl fluoride and trifluroacetyl fluoride (in the presence of water vapor, the carbonyl fluoride rapidly converted into HF and CO_2).

Warity and Kwan (1968) have reported work which indicates that the principal toxicants evolved from PTFE decomposition are absorbed to the particulates generated by combustion. The major toxicant, at temperatures above 450°C, was reported to be PFIB (perfluoroisobutylene), for which the LC_{50} (of the pure gas) was reported to be 40,000 ppm, much less lethal than the decomposition products, which calls this conclusion to question. In 1982, however, Smith *et al.*, evaluated PFIB for LC_{50}'s based on a series of short term exposures. They found a 15 second LC_{50} of 361 ppm and a ten minute LC_{50} of 17 ppm. These results would seem more creditable and suggest PFIB is a real hazard.

Coleman *et al.*, (1968) reported that between 550 and 600°C, the proportion of carbonyl fluoride in PTEF decomposition streams ran as high as 63%. Scheel *et al.*, (1968) have reported a 24-hour LC_{50} for carbonyl fluoride of 360 ppm, not sufficient to account for the lethality seen with PTFE decomposition. Histopathological changes in liver and lung were reported to be associated with CoF_2 lethality.

Liang *et al.*, (1980) reported a 90-day study on the exposure of rabbits (three hours per day) to PTFE degradation products. The products were identified as hexafluoropropylene (1.0-1.2 mg/m^3), tetrafluoroethylene (6-8 mg/m^3), octafluorocyclobutane (12-16 mg/m^3), hydrogen fluoride, carbonyl fluoride (0.3 mg/m^3), and a trace amount (0.005 mg/m^3) of perfluoroisobutylene. Reported effects included primarily damage to the lungs.

Grupp (1975) showed that environmental humidity and the degree of corrosion to the upper respiratory tract resulting from PTFE exposure were directly proportional, as one would expect with acid gases.

Kaplan *et al.*, (1984) evaluated five halogenated polymers in the NBS test system. These were two chlorofluoropolymers, Halar 500 (ethylene-chlorotrifluoro ethylene) and Halar 555® (blown ethylene-chlorotrifluoroethylene) and three fluoropolymers, Teflon 100® (fluorinated ethylene/fluorinated propylene), Tefzel 200® (ethylene-tetrafluoroethylene) and Kynar® (polyvinylidene fluoride). The halogen content of these materials were: Halar 500® and Halar 555®, 39.4% fluorine and 24.6% chlorine; Teflon 100®, 76.0% fluorine; Tefzel 200®, 59.4% fluorine; and Kynar®, 59.4% fluorine. Levels of HF, HCL and COF_2 (carbonyl fluoride) evolved were measured in addition to the standard measures.

The smoke produced by Halar 500®, Halar 555® and Kynar® exhibited comparable toxicities in both combustion modes, with LC_{50} values that ranged from 15.1 to 33.3 mg/L. Nonflaming combustion produced high levels of CO and was associated with high incidences of incapacitation and lethality during exposure. under flaming combustion, small quantities of CO were generated and low incidences of incapacitation and exposure lethality were seen. Lethalities primarily occurred post exposure and were possibly related to the HF and/or HCl evolved. With Teflon 100®, LC_{50} values in flaming (0.08 mg/L) and nonflaming (<0.05 mg/L) modes were comparable, but toxic effects differed considerably. With Tefzel 200, differences between combustion modes were observed both in toxicity (LC_{50}'s were, flaming 30.2, nonflaming 3.3 mg/L) and in toxic effects. With both Teflon 100® and Tefzel 200®, levels of CO, HF, HCl and COF_2 could not account for the observed effects. The results demonstrate that combustion atmospheres generated by decomposition of halogenated materials at different temperatures may differ considerably in toxicological effects.

The actual toxicities of the decomposition products, and the reasons for the sharp division in degree of toxicity between the members of this class, have also been widely published and much debated. Epstein and Heicklen (1969) reported comparative toxicities of carboxynitroso rubber in the following series < PEP polymer < Viton (a fluorelastomer) < polyamide film < FEP < PTFE, but did not quantitate how much more toxic PTFE was than the other members of this series.

Harris *et al.*, (1973) reported that sublethal exposures of rats to fluorinated polymer thermolysis products caused marked diuresis and increased water consumption, increased urinary fluoride levels, glucosuria, and an increase in urinary potassium. Along the same lines, Polakoff *et al.*, (1974) has reported increased urinary fluoride levels in PTFE fabricators, perhaps due to their metabolism of carbonyl fluoride.

Gregory (1973) reported the comparative toxicities to be (highest to lowest) FEP > Viton > Fluorel (a fully saturated fluorinated polymer containing more than 60% fluorine by weight) > a PFP + vinylidene fluoride polymer.

Scheel, McMillan and Phipps (1968) described the toxicity syndrome related to daily inhalation of sublethal doses of PTFE as resembling fluorosis. They found that there was a direct correlation between PTFE exposure and urinary fluoride levels.

Cater *et al.*, (1974) reported no correlation between hydrolyzable fluoride levels in exposures (to decomposition products of four fluoropolymers) and the lethality seen within 24 hours of exposure, indicating that the toxicity is not purely a matter of the effects of fluoride. It is not clear, however, that this is the case with the nonlethal effect seen with exposure to PTFE heated to temperatures of 350°C (*i.e.*, polymer fume fever).

Table 7-5 summarizes reported quantitative combustion toxicity evaluations of PTFE in each of several systems.

Table 7-5. Combustion Toxicity of PTFE

Reference	Condition/System	Reported Result	
Alaire and Anderson (1979)	Univ. of Pittsburgh system		LC_{50} = 0.64 mg/L LT_{50} = 8 minutes
Young (1976)	Pure Gas		LC_{50} = 40 mg/L
Levin *et al.*, (1982)	NBS	Flaming: Nonflaming:	LC_{50} = 0.3 - 2.6 LC_{50} = 0.9 - 5.02
(Previously unpublished data, S.C. Gad)	CMIR, 40°/minute	Flaming: Nonflaming:	LC_{50} = 0.05 mg/L LC_{50} = 0.70 mg/L

NBS = National Bureau of Standards
CMIR = Carnegie Mellon Institute of Research (Union Carbide)
NBS and CMIR system results are directly comparable. None of the other system results are readily intracomparable.

Table 7-6 is a similar summary for fluoropolymers other than PTFE. As can be readily seen, PTFE and FEP are markedly more toxic than any of the other fluoropolymers or than any of the other plastics evaluated to date. The remaining fluoropolymers (other than PTFE and FEP), on the other hand, are not significantly different from wood in the toxicity of their thermolysis products. Again, atmospheres produced at different temperatures (heating rates) did differ in their toxicological effect.

POLYAMIDES AND POLYIMIDES

Though a very significant class of polymers in terms of both combustion and exposure potential, the polyamides have not been studied extensively in the context of combustion toxicity (though they were among the first materials studied).

In 1962, MacFarland and Leong characterized the response of animals exposed to nylon smokes as a slowly developing pulmonary edema which caused death. The associated LCT_{50} [2] was 1.9 × 106 mg/min/m^3 (which, for a thirty minute exposure, would translate to 63 mg/L). In 1975, Bulzgin observed that the polyamides have a small oxygen to nitrogen ratio, and therefore are more thermostable than most polymers, but should also be more likely to generate HCN. The HCN generated, however, appears to be absorbed

2 The LCT50 is the timed/concentration product which produces death in 50% of exposed animals.

Table 7-6. Combustion toxicity of fluoropolymers other than PTFE

Material	Reference	Condition/System	
Polyvinyl fluoride (PVF)	Kourtides *et al.*, 1978	USF	ALC_{50} = 65.9 ± 9.44 mg/L
	Young *et al.*, 1976	USF	LC_{50} = 56 ± 3 mg/L
Polyvinylidine fluoride	Kourtides *et al.*, 1978	USF	ALC_{50} = 98.60 ± 7.43 mg/L
	Kaplan *et al.*, 1984	NBS	Flaming : LC_{50} = 27.3 mg/L Nonflaming: LC_{50} = 24.3 mg/L 440°C LC_{50} > 42.5 mg/L
Chloro-fluoroethylene (CTFE)	Birnbaum, Scheel and Coleman, 1969	Author's own system, 3 hr. LC_{50}s	@ 373°C = 31.5 g/L @ 400°C = 23.5 g/L
Ethylene chlorotri-luoroethylene (ECTFE)	Scheel *et al.*, 1976	Author's own system, 1 hr	LC_{50} = 42.5 mg/L
ETFE	Kaplan *et al.*, 1984	NBS	Flaming: LC_{50} = 20.0 mg/L Nonflaming: LC_{50} = 27.8 mg/L 440°C LC_{50} = 33.3 mg/L
Fluorinated ethylene and propylene (FEP)	Kaplan *et al.*, 1984	NBS	Flaming: LC_{50} = 0.075 mg/L Nonflaming: LC_{50} = 0.05 mg/L 440°C 1 of 6 dead at 1 mg/L
Ethylene tetrafluoro-ethylene	Kaplan *et al.*, 1984	NBS	Flaming: LC_{50} = 30.2 mg/L Nonflaming: LC_{50} = 3.33 mg/L 440°C LC_{50} > 3.3 mg/L

USF = University of San Francisco
ALC_{50} = Apparent LC_{50} (animals not followed out beyond day of exposure)

to the solid phase of the smoke. Terrill *et al.* (1978) also reported that thermolysis of nylon produced significant amounts of NH_3 and HCN.

Yoshida *et al.*, (1978) decomposed both Nylon 6 and 66 in room air at 550°C, generating an atmosphere consisting primarily of acetaldehyde, carbon monoxide and ammonia, but also containing carbon dioxide, methane, ethylene, propylene, benzene and particulates which were primarily over 100 μm in diameter (and therefore would not be respirable). This decomposition stream (in a flow through system) was determined to have a one hour ALC_{50} (apparent lethal concentration 50%) in mice of 79.6 mg/L.

It should be noted that this was a flow through (dynamic) system and therefore the LC_{50} is not comparable to those determined by most other systems (which are static).

Young *et al.*, (1976) reported LC_{50} data in a closed system for three polyamide forms; a "plain" polyamide (LC_{50} = 32 mg/L), a chlorinated polyamide (LC_{50} = 22 mg/L) and a fluorinated polyamide (LC_{50} = 7 mg/L).

Kourtides *et al.*, (1978) reported an ALC_{50} of 10.75 ± 2.25 mg/L. Kishitani and Nakamura (1974) reported a study in which they evaluated both wood (LC_{50} = 9.7 mg/L) and nylon (LC_{50} = 1.0 mg/L).

Gad and Kaplan (1984) reported extensive evaluations of Nylon 6 in both the NBS and CMIR test systems. In addition to the standard measures, incapacitation (by leg flexion) and carboxy-hemoglobin levels were also measured in the NBS tests.

In the NBS system, the LC_{50} value for flaming combustion of Nylon 6 was determined, by probit analysis, to be 83.3 mg/L. In two replicate experiments a material charge (total sample loaded into the combustion device and assumed to be totally decomposed) of 90 mg/L produced 60 percent and 40 percent lethalities and 99 mg/L resulted in 100 percent lethality. In two other experiments, lethalities did not occur with a charge of 50 mg/L of sample; however, a material charge of 120 mg/L caused 60 percent lethalities. In the latter experiment, sample overloading resulted in an explosion and leakage of combustion atmosphere from the chamber (CO concentration decreased), invalidating the data.

The predominant toxic effects produced by flaming combustion atmospheres of Nylon 6 were incapacitation and post exposure lethality. Four and five animals were incapacitated in one 90 mg/L experiment and in a 99 mg/L experiment, respectively. Although lethalities occurred primarily post-exposure, one animal died during exposure in each of two experiments. The occurrences of incapacitating effects were, in general, consistent with the analytical measurements of CO in the combustion atmospheres. The data showed that, at approximately 200 ppm CO, incapacitation may be anticipated after approximately 17 minutes of exposure (C•t product of 34,000 ppm-min).

However, the blood COHb levels of incapacitated animals were somewhat below the level of 70 percent COHb usually measured in animals incapacitated by 2000 ppm CO. Also, the COHb levels in these incapacitated animals and in the few animals that died during exposure were less that would be anticipated from the C•t products calculated for CO in these experiments. One possible explanation is that irritant may be generated by flaming combustion of the material. This irritant could have inhibited the respiration rate of animals and delayed loading of CO by the animals. In addition, the irritant may have contributed to the effect of CO on incapacitation and the limited exposure lethalities. The presence of an irritant effect is further suggested by both the high incidence of post exposure lethalities and the clinical signs observed post exposure. However, it should be noted that heat spikes, elevated chamber temperatures, elevated levels of CO_2 and oxygen depletion occurred during these flaming combustion experiments.

The NBS LC_{50} value for the nonflaming combustion of Nylon 6 was determined by probit analysis to be 110.1 mg/L. Five experiments with sample concentrations of 29.8 mg/L to 100 mg/L were completed, with only one lethality being observed in each of two experiments and two lethalities in the highest sample experiment. Incapacitation of animals was not observed except in the high dose experiment and, with one exception, the few deaths that did occur in these experiments were post exposure lethalities. Much less carbon monoxide was generated by nonflaming than flaming combustion, as was evident from a comparison of maximum CO concentration and C•t products in the two modes. The low CO and COHb analytical measurements are consistent with the low incidence of incapacitation and exposure lethalities observed in these experiments. Post exposure lethalities, which occurred only at the highest sample concentrations and in limited number, are often associated with the presence of an irritant. However, the low incidence of lethalities during and after exposure are not indicative of a strong irritant.

Exposure of animals to a nonflaming Nylon 6 combustion atmosphere of produced a few symptoms indicative of exposure to an irritant gas. However, the symptomatology was much less severe than that produced by flaming combustion atmospheres. Leg-flexion responses, vocalizations and struggling movements, beginning a few minutes into the exposure, were characteristic symptoms at the beginning of exposure. These symptoms subsided during the 30-minute exposure period. At the termination of exposure, the muzzles of the animals were typically wet and, in two experiments, respiratory symptoms of congested breathing or rapid breathing were observed. The few post exposure lethalities occurred within 24 hours of exposure.

The NBS LC_{50} value for flaming combustion of Douglas fir in concurrent studies was determined to be 37.6 mg/L. The predominant toxic effects produced by this mode of combustion were incapacitation and exposure lethalities. Post exposure lethalities did not occur in any of the experiments. These effects were consistent with the COHb levels measured at incapacitation (68.9 - 80.5 percent). The occurrence of incapacitation and exposure lethalities and the measured COHb blood levels were also consistent with the analytical data for CO in the combustion atmospheres (*i.e.*, C•t products at 93,900 - 172,740 ppm-min.) The COHb and CO analytical data demonstrate that the observed toxic effects were caused by the large quantities of CO evolved by Douglas fir.

In the CMIR system, LC_{50} values (in mg/L) were determined to be as follows:

	20°C/MINUTE HEATING RATE	30°C/MINUTE HEATING RATE	40°C/MINUTE HEATING RATE	50°C/MINUTE HEATING RATE
NYLON 6	69.4	57.2	56.8	36.8
Trials (N)	6	5	5	6
DOUGLAS FIR	43.8	38.5	39.3	24.6
Trials (N)	6	6	6	6

Thus, it appears that the polyamides stand out as being in need of further evaluation. It seems clear that they are worse than woods, but perhaps (due to particulates "capturing" some of the more toxic materials in a nonrespirable phase) not significantly worse. One cannot presently relate these to other polymers.

There is only one report in the literature on polyimides. Epstein and Heicklen (1969) reported that at temperatures between 500 and 700°F, the decomposition streams from polyimides were more toxic than those from PTFE. it should be remembered that in this temperature range there is minimal decomposition of PTFE.

PHENOLICS

This is another group of materials on which a separate review has been conducted (Johnston *et al.*, 1988). Hilado and Huttlinger (1978) reported having investigated some materials of the class but gave insufficient information to be able to evaluate their results. Anderson *et al.*, (1979) compared the decomposition products of urea formaldehyde and phenol formaldehyde in the University of Pittsburgh system, and determined that urea formaldehyde was the worst of the two and that it acted by inducing a cardiomyopathy. Alaire and Anderson (1979) also reported that in the Pittsburgh system, phenol formaldehyde had an LC_{50} of 6.3 mg/l and an LT_{50} of 20 minutes. These results are sufficient to be able to place the phenolics as being significantly worse than wood but significantly better than PTFE. Also reported was a strong sensory irritant component associated with the phenolics which probably reflective of formaldehyde being generated as a decomposition product. Further evaluation is called for if a more exact evaluation is required. Furthermore, the cardiomyopathy finding is an interesting (and for this type of work, unusual) one and should be pursued further.

In the FAA system, Crane *et al.*, (1977) found times to incapacitation of from 3.07 to 12.26 minutes and times to deaths of from 5.43 to18.22 minutes.

POLYCARBONATES

Kourtides *et al.*, (1978) reported an ALC_{50} for bisphenol-A-polycarbonate of 15.0 ± 4.33 mg/L. Kimmerle and Prager (1980) reported that to achieve the same LC_{50} for polycarbonate as for cellulose or woods requires higher temperatures - that is, that the polycarbonates decompose at higher temperatures than wood or cellulosics. In their system, polycarbonates had an LC_{50} of from 26 to 36 mg/L (as compared to that of 29 mg/L for wood), making this class not at all different from wood in combustion toxicity.

Sangha *et al.*, (1981) have reported evaluations of a series of nine polycarbonates in the university of Pittsburgh system (which assigns wood an LC_{50} of 63.8g). LC_{50} values ranged from 8.3 to 22.7 grams, with seven of the nine having values less than 15 grams. This is four times more toxic than wood.

POLYETHYLENES

The polyethylenes as a class have been briefly assessed by a number of researchers. One issue which emerges from the papers reviewed below is the need for better sample identification - few researchers have bothered to identify the materials they study as either high or low density (a difference which should be important).

Michael *et al.*, (1976) performed a detailed analysis of the products of pyrolysis, "thermooxidation", and combustion and determined most of the products to be CO_2, CO, or C_3 to C_{15} aldehydes. The numerical toxicity data reported to date are summarized in Table 7-7.

Polyethylenes thus appear to be no worse than wood - in fact, there are some indications that they may be less toxic than wood.

POLYPROPYLENES

Michael *et al.*, (1979) reported that the analysis of thermolysis stream products from polypropylene gave similar results to those for polyethylene. Woolley *et al.*, (1979) determined that the pyrolysis products of polypropylenes consisted primarily of saturated aliphatics, while the combustion products were a mixture of simple and complex aldehydes and ketones. Kishitani and Nakamuera (1976) reported that the resulting toxicity from decomposing polypropylenes could be accounted for by the carbon monoxide produced.

Table 7-7. Combustion Toxicity of Polyethylenes

Type	Reference	Condition/ System	Reported Result
NS	Kourtides, 1978	USF	ALC_{50} = 11.81 ± 3.73 mg/L
LD	Hilado and Brauer, 1979	USF	T_D 11.67 min T_i 9.93 min
HD	Gad and Smith, (1983)	CMIR Nonflaming	LC_{50} 25.49 mg/L T_i 8.13 min
		Flaming	LC_{50} 58.82 mg/L T_i 7.49 min
LD	Hilado and Solis, 1977	USF	T_D 22.60 ± 0.62 T_i 16.68 ± 2.23

NS = Not Specified
LD = Low density
HD = High density
T_D = Time to death
T_i = Time to incapacition

Cornish (1973) reported an LC_{50} value for polypropylene of 17 mg/L in the University of Michigan system and that dynamic mode (in which more air is present) combustion produced move toxic components.

Hilado *et al.*, (1979) reported toxicity (in the USF system) to be comparable to, or less than, that of the polyethylenes, and that it (with a T_i of 15.31 minutes and a T_d of 21.4 minutes) was primarily a carbon monoxide producer.

Purohit and Orzel have published a separate review of the literature on this class.

POLYSTYRENES

Pfaffli and Zitting (1978) investigated the components of polystyrene decomposition streams at 200, 300, and 500°C. They found that as the temperature was increased, more short chain aromatic hydrocarbons were generated (at 200°C, the product of decomposition was almost entirely styrene). In a related study, Zitting *et al.*, (1978) reported that mice exposed to these decomposition streams displayed increased levels of liver microsomal enzymes and glutathione. Supporting this result is a report by Heimann (1979) that such exposures to decomposition streams caused temporary increases in liver weight to body weight ratios of exposed animals - another indicator of liver enzyme induction. Results are summarized in Table 7-8.

Table 7-8. Combustion Toxicity of Polystyrenes (PS)

Reference	Condition/System	Reported Result		
Hilado and Solis, 1977	USF	TD 26.16 + 0.12		
		T_I 19.04 + 0.39		
Alaire and Anderson,	University of Pittsburgh	Expanded PS		-LC_{50} 5.8 mg/L
				-LT_{50} 11 min
		Expanded PS + Flame Retardant		-LC_{50} 10.0 mg/L
				-LT_{50} 9 min
Previously unpublished (S. C. Gad)	NBS	Expanded PS	Nonflaming	-LC_{50} 40 mg/L
				-EC_{50} 27 mg/L
			Flaming	-LC_{50} 24 mg/L
				-EC_{50} 19 mg/L
Levin *et al.*,	NBS		Nonflaming	-LC_{50} 40-50 mg/L
			Flaming	-LC_{50} 33-53.5 mg/L

In summary, polystyrene decomposition products appear acutely to be no different than wood. Long term toxicity effects may well be a different matter.

SILICONE POLYMERS

Only one study (Hilado *et al.*, 1978) has been reported for this class. In this report the silicone polymer tested was reported to generate CO as its principal component. However, the products of this polymer appeared to be the least toxic of all materials evaluated under the conditions of test. Further evaluation in another system would appear to be warranted.

URETHANES

Because of the valid expectation that thermally decomposing urethanes would act as significant sources of hydrogen cyanide gas, this class has been very widely evaluated.

The nature of the toxicant gases generated - and the factors influencing their generation - has turned out to be more complicated than was once thought.

The toxicity of urethane decomposition products is particularly sensitive to the construction method of the plastic (Hoffman and Sand, 1974), the decomposition temperatures, and the catalysts utilized in the polymerization process. Construction alters both the chemical and physical nature of the decomposition products. With some structures one gets denser smoke consisting of more respirable particle sizes (and as Bulzgin, 1975, reported, much of the HCN generated by urethanes absorbs onto the solid phase of the smoke that is generated). Decomposition products include isocyanates (such as toluene diisocyanate, from polymers where unreacted isocyanate catalysts remained; Kimmerle and Prager, 1980; Terril *et al.*, 1978), CO, HCN, CO_2, phosgene, HCl (Paulson and Moran, 1974; Tatem and Williams, 1975), and isocyanic acid esters (Seemann and Woelcke, 1976). In some urea molding resins, sulfur dioxide and H_2S are also produced. Jellinek and Takada (1977) explored the question of the effects of polymer density and surface area on HCN evolution rates, and determined that neither had a significant effect - that the rate was 2.1×10^6 mg of HCN per minute per cubic meter of air passed through a decomposition system. Sumitomo (1974) attempted (without notable success) to relate the combustion toxicity to the total nitrogen content of the polymer.

The range of mechanisms of toxicity seen with the urethanes supports the belief that the decomposition products are a mixture with several different biological actions. MacFarland and Leong (1962) reported that all rat deaths occurred within three days of exposure - apparently by obstructive occlusion of the lung (it should be kept in mind that neither CO or HCN would take up to three days to act). Alaire *et al.*, (1975) first pointed out that there was a strong sensory irritant effect associated with exposure to the urethane decomposition stream. Hartung *et al.*, (1977) evaluated urethanes in terms of their actions on rotorod performance in rats (as compared to the action of wood), and found that polyurethanes were significantly less toxic (in terms of how fast they acted to incapacitate animals) than wood. McGuire and Annau (1980) used an NBS system with attachments to measure alterations in licking behavior (and through it to quantitate the irritant response), and shock avoidance (which is primarily a response to CO and HCN). The results led them to propose that the incapacitation aspects of the response to urethane decomposition streams is a two phase effect (that is, that incapacitation is by two independent mechanisms which are affected differentially).

Numerical data are reported in Table 7-9. It appears that the urethanes are in the same class as the phenolics and cellulosics - significantly more lethal than woods, but not to the extent that PTFE is. In terms of their potential to incapacitate, the decomposition streams from the urethanes are seemingly not at all different from those of the woods.

Table 7-9. Combustion toxicity of urethane

Material	Reference	Condition/System	Reported result
Polyurethane	MacFarland and Leong, 1962	Author's own	LC•t = 50 mg/min/m^3
Polyurethane (GM21)	Previously unpublished data, S. C. Gad	NBS	Nonflaming: LC_{50} = 14 mg/L EC_{50} = 20 mg/L Flaming: LC_{50} = 42 mg/L EC_{50} = 10 mg/L
Polyurethane foams (10 different)	Parent *et al.*, 1979	Author's own	LC_{50} = 20.2 - 53.0 mg/L
Flexible polyurethane foam (GM21)	Alaire and Anderson, 1979	University of Pittsburgh	LC_{50} = 13.0 mg/L LT_{50} = 13 minutes
High resilience polyurethane foam (GM25)	Same	Same	LC_{50} = 8.3 mg/L LT_{50} = 19 minutes
Rigid polyurethane foam (GM29)	Same	Same	LC_{50} = 10.4 mg/L LT_{50} = 28 minutes
Rigid polyurethane foam, fluorocarbon blown (GM35)	Same	Same	LC_{50} = 7.5 mg/L LT_{50} = 17 minutes
Rigid polyurethane foam, CO_2 blown (GM37)	Same	Same	LC_{50} = 8.0 mg/L LT_{50} = 15 minutes
Rigid isocyanate foam (GM41)	Same	Same	LC_{50} = 6.4 mg/L LT_{50} = 18 minutes
Urea formaldehyde foam	Same	Same	LC_{50} = 2.5 mg/L LT_{50} = 22 minutes
Flexible polyurethane	Levin *et al.*, 1982	NBS	Nonflaming: LC_{50} = 37.9 - 50.9 mg/L Flaming: LC_{50} = 38.0 - 49.5 mg/L

VINYLS

The polyvinyl chlorides have been extensively evaluated in terms of their combustion toxicity (Stark, 1969; Saito, 1970; Smith, 1972; O'Mara 1974; Michael, 1976), probably due to the (correct) assumption that the material would generate a lot of acid gases upon decomposition.

A considerable amount of work has been done to establish the nature of the decomposition products generated. Colardyn *et al.*, (1976) reported that combustion of PVC generates HCl, CO_2, CO, and very small quantities of phosgene, making the resulting mixture primarily the cause of an irritant gas problem (a conclusion reinforced by a study of exposures of 20 people - in all of whom an acute obstructive lung disease was seen). Einhorn (1977) reported that much particulate was also generated, with the HCl and phosgene absorbed to it (and with animals that were exposed to the decomposition products showing respiratory and metabolic acidosis).

Terrill *et al.*, (1978) investigated the effects of altering airflow and decomposition temperature on the amounts of various decomposition products. CO_2 increased with increased temperature or airflow. CO decreased with increased airflow and increased with temperature. Benzene decreased with both airflow and temperature. Methane decreased with airflow and temperature. HCl was apparently independent of both airflow and temperature. Other products, such as ethane, propylene, and vinyl chloride, were produced in very small quantities. The relative concentrations produced were in the order CO_2 > HCl > CO > benzene > methane.

Brown and Birkey (1980) reported determinations of carbonyl chloride (phosgene) levels produced from PVC when it was thermally decomposed under four different sets of conditions. Herpol (1980) reported the comparative levels of CO_2 concentrations produced by identical weights of wood and three different PVC forms (rigid, flexible and a foam) at three different temperatures, in which only one of the three forms was found to produce amounts of CO_2 comparable to wood (this being flexible PVC at 800°C, at which temperature the material releases almost four times as much CO_2 as wood).

Boettner and Ball (1980) evaluated the levels of several materials generated by the hot wire cutting of PVC film (as done in supermarkets for the wrapping of meats). The total quantities, generated as vapors, for HCl, plasticizer, benzene, toluene, acrolein and CO were found to be in the μg to ng range for each cut of film.

Table 7-10. Combustion Toxicity Of Vinyl Polymer

Material	References	System	Results
Polyvinyl Chloride(PVC)	Kourtides, *et al.*	USF	ALC_{50} = 34.20 ± 1.0 mg/L
PVC	Hinderer, 1984	NBS	LC_{50} = 17.3 mg/L
PVC (46% homopolymer)	Alarie and Anderson (1979)	University of Pittsburgh	LC_{50} = 15.2 grams
(92% homopolymer)			LC_{50} = 7.0 grams LT_{50} = 10 minutes
PVC	Sangha *et al.*, 1981	University of Pittsburgh	LC_{50} = 13.3 grams
PVC	Cornish *et al.*, 1975	University of Michigan	LC_{50} = 2.0 grams (where wood = 3.2)
PVC	Hilado and Cummings, 1978	USF	T_i = 5.95 - 12.69 T_d = 16.3 - 16.88

Tsuchiya and Sumi (1967) also reported extensive analytical evaluations of the decomposition products of PVC. The PVC evaluated was a powdered resin without plasticizer, which removed somewhat the case being evaluated from the actual case of interest. The evaluations were performed with varied decomposition conditions (temperature and air). In an oxygen depleted situation, HCl is markedly the predominant product. The data indicate that in low oxygen situations, there is less production of HCl than at the highest oxygen levels evaluated. Also, as one would expect, HCl generation levels decreased at higher temperatures.

The numerical data as to toxicity are summarized in Table 7-10.

In an evaluation of the pulmonary effects of PVC thermal decomposition products in guinea pigs, Wong *et al.*, (1983) found that the HCl generated caused extensive damage to the upper respiratory tract that could be detected functionally until 57 days after exposure.

Hinderer (1984) has published a separate review of PVC combustion toxicity which does a good job of reviewing the available information.

MISCELLANEOUS MINOR POLYMER FAMILIES

These classes are minor, of course, only in the sense that each has been evaluated, in the context of combustion toxicity, in a minor way. As a result, reaching a balanced final evaluation for any of these materials is difficult.

This is not to say that the data which exists is not good, and in the interest of trying to offer a current overview which is as complete as possible, a summary is presented below.

Latex

Sharrod and Edser (1976) evaluated latex in the form of textiles, reporting that the only major products generated were CO and smoke.

Polychloroprene (PCP/neoprene)

Hilado (1976) reported an evaluation of PCP at two different temperatures (500 and 800°C), with the added variable that the airflow was low at the lower temperature and high at the upper temperature. In both cases the material was evaluated relative to polyurethane, being determined to be "worse" at the lower temperature and "not as bad" at the higher.

Alarie and Anderson (1981) also reported an evaluation of PCP, but as their sample included 5% zinc ferrocyanide as a flame retardant, these data are presented in the later section on flame retardants.

Polyacrylonitrile

Both reports found in the literature seem to make it clear that this material is one which generates significantly more toxic products than others. Kishitani and Nakamura (1974) evaluated a polyacrylonitrile in their dynamic exposure system (in which wood in the flaming mode had an LC_{50} of 9.3 mg/L). In the nonflaming mode, polyacrylonitrile had an LC_{50} of 0.32 mg/l, while in the flaming mode its LC_{50} was 0.5 mg/l. Woolley *et al.*, (1979) reported that these generated more HCN than did polyurethanes (in fact, some 36% of the total sample weight ended up as HCN).

Polyether Sulfone

Young *et al.*, (1976), in the USF system, reported an LC_{50} for polyether sulfone of 55 mg/L (with PTFE being determined to be 65 mg/L at the same time). Hilado and Solis (1977), for the material in the same system, reported a TD of 5.50 ± 0.28 and a T_i of 300 ± 0.81, making this the most toxic of all the materials evaluated by these researchers. Evaluation of these materials in a static system should be a priority concern. Sangha *et al.*, 1981, using the University of Pittsburgh system, determined an LC_{50} of 14.40 grams (about a quarter that of wood).

Polyphosphazene

In two separate publications, Lieu, Magill and Alaire (1979) reported evaluations of copolymers of the polyphosphazenes. The materials were found to generate particularly lethal products. It was determined that the results were not due to HCN or CO, but rather to some other, more lethal, gas or gases of an undefined nature. This class also merits some priority in further evaluation.

Polyphenylene Oxide

Ball, Weiss and Boethner (1970) reported the analytical determination of the nature of the products generated by decomposition of this polymer. It was found to generate mostly CO_2 and CO (~ 40% of the total), plus methane, toluene, ethylbenzene, styrene (each of these in the range of from 10 to 40 mg per gram of material), benzene, straight chain aliphatics, methanol and acetaldehyde (this last group in the range of from 0.1 to 4 mg per gram of sample). No toxicity evaluation (as such) has been performed.

Sangha *et al.*, using the University of Pittsburgh system, determined an LC_{50} for a polyphenylene oxide of 13.15 g (less than a quarter that of wood).

REFERENCES

Alarie, Y. C. and Anderson, R. C. (1979) Toxicologic and Acute Lethal Hazard Evaluation of Thermal Decomposition Products of Synthetic and Natural Polymers. *Toxicol. Appl. Pharmacol.* **51**: 341-362.

Alarie, Y. C., Anderson, R. C. (1981) Toxicologic classification of thermal decomposition products of synthetic and natural polymers. *Toxicol. Appl. Pharmacol.* **57**: 181-88.

Alarie, Y.C., Stock, M.F., Matijah-Schaper M. and Birky, M.M (1983) Toxicity of Smoke During Chair Smoldering Tests and Small Scale Tests Using the Same Materials, *Fund. Appl. Toxicol.* 3:619-626.

Alarie, Y. C., Wilson, E., Civic, T., Magill, J. H., Funt, J. M., Barrow C., and Frohliger, J. B. (1975) Sensory Irritation Evoked by Polyurethane Decomposition Products. *Combust. Toxicol.* **2**: 139-150.

Anderson, R. C. and Alarie, Y. C. (1978) Screening Procedure to Recognize "Supertoxic" Decomposition Products From Polymeric Materials Under Thermal Stress. *Journal of Combustion Toxicol.* **5**: 54-63.

Anderson, R. C., Stock, M. F., Sawin, R. and Alaire, Y.,C. (1979) Toxicity of Thermal Decomposition Products of Urea Formaldehyde and Phenol Formaldehyde Foams. *Toxicol. Appl. Pharmacol.* **51**: 9-17.

Arito, H. and Soda, R.(1979) Pyrolysis Products of Polytetrafluoroethylene and Polyfluoroethylenepropylene with Reference to Inhalation Toxicity. *Ann. Occup. Hyg.* **20**: 247-255.

Ball, G., Weiss, B., Boettner, E.,A. (1970) Analysis of the Volatile Combustion Products of Poly(Phenylene Oxide) Plastics. *Amer. Ind. Hyg. Ass. J.* **31**: 572-8.

Benbow, A. W. and Cullis, C. F. (1975) The Formation of Toxic Products During the Combustion of Halogen - Containing Polymers. *Eur. Polym. J.* **11**: 723-727.

Birky, M., Paabo, M., Levin, B. C., Womble, S. E., and Malek, D. (1980) *Development of Recommended Test Method for Toxicological Assessment of Inhaled Combustion Products.* National Bureau of Standards, U.S. Department of Commerce. NBSIR 80-2077.

Birnbaum, H. A., Scheel, L. D., Coleman W. E. (1968) Toxicology of the Pyrolysis Products of Polychlorotrifluoroethylene. *Amer. Ind. Hyg. Assoc. J.* **29**: 61-5.

Blum, A. (1977) Flame Resistant Additives as Possible Cancer Hazards. *Science* **195**: 17-23.

Boettner, E. A. and Ball, G. L. (1980) Thermal Degradation Products From PVC film in Food-wrapping Operations. *Am. Ind. Hyg. Assoc. J.* **41**: 513-522.

Brown, J. E. and Birky, M. M. (1980) Phosgene in the Thermal Decomposition Products of Poly(vinyl chloride): Generation, Detection and Measurement. *J. Anal. Toxicol.* **4**: 166-174.

Bulygin, B. M. (1975) Thermal Decomposition of Nitrogen Containing Polymers. *Plast. Massy.* 5:24-27.

Carter, V. L., Bafus, D.,A., Warrington, H.,P. and Harris, E.S. (1974) The Acute Inhalation Toxicity in Rats From the Pyrolysis Products of Four Fluoropolymers. *Toxicol. Appl. Pharmacol* **30**: 369-376.

Colardyn, F., Van der Straeten, M., Lamont, H. and Van Peteghem, T. (1976) Acute Inhalation Intoxication by Combustion of Polyvinylchloride. *Int. Arch. Occup. Environ. Hlth.* **38**: 121-127.

Coleman, W. E., Scheel, L.D., Kupel, R. E., Larkin, R.,L., (1968) Identification of Toxic Compound in the Pyrolysis Products of Poly (tertrafluoroethylene). *Amer. Ind. Hyg. Assoc. J.* **29**: 33-40.

Cornish, H. (1973) Toxicity of Thermal Degradation Products of Plastics. **In**: *Proceedings of Symposium on Products of Combustion of (Plastic) Building Materials*. Lancaster, PA, March 26-27, pp. 30-33.

Cornish, H. H., Hahn, K. J., and Barth, M. L. (1975) Experimental Toxicology of Pyrolysis and Combustion Hazards. *Environ. Health. Perspect.* **11**: 191-196.

Crane, C. R., Sanders, D. C., Endecott, B. R., Abbott, J. K., and Smith, P. W. (1977). *Inhalation Toxicology: I. Design of a Small Animal Test System. II. Determination of the Relative Toxic Hazards of 75 Aircraft Cabin Materials.* Prepared for U.S. Department of Transportation, Washington, DC. Report No. FAA-AM-77-9.

Einhorn, I. N. (1977) Methodology for the Study of Toxicology in Combustion. Application to PUC. *J. Macromol. Sci. Chem.* **8**: 1519-28.

Epstein, G. and Heicklen, J. (1969) Relative Toxicity of Selected Polymeric Materials Due to Thermal Degradation TR-0066 17p.

Gad, S. C., and Kaplan, H. (1984) *Combustion Toxicity Studies of Nylon 6. An Industry Approach to Chemical Risk Assessment.* Industrial Health Foundation, pp. 250-263.

Gad, S. C. and Smith, A. C. (1983) Influence of Heating Rates on the Toxicity of Evolved Combustion Products: Results and a System for Research. *J. Fire Sci.* **1**: 465-479.

Gregory, Arthur R. (1973) Toxicity of Thermolysis Products of Spacecraft Elastomers. *J. Appl. Polym. Sci.* **17**: 2909-10.

Grupp, H. (1975) Corrosive Attack of Pyrolytic Gases Containing Hydrogen Chloride as Influenced by Air Humidity. *Maschineschaden* **48**: 119-124.

Harris E. S., Warrington, H. P., Bailey, J. W., Dilley, J.V. (1973) Response of Rats to Pyrolysis Products of Fluorinated Polymers *Toxicol. Appl. Pharmacol.* **25**: 470-479.

Hartung, R., Ball, G. L., Boettner, E. A., Rosenbaum, R., and Hollingsworth, Z. R. (1977) The Performance of Rats on a Rotarod During Exposure to Combustion Products of Rigid Polyurethane Foams and Wood. *J. Combust. Toxicol.* **4**: 506-522.

Heimann, R., Skornik, W., and Jaeger, R. J. (1979) Biochemical Changes in Guinea Pig Lungs Following Exposure to Plastic Combustion Products, Paraquat, or ANTU. *Toxicol. Appl. Pharmacol.* **48: A58.**

Herpol, C. (1980) Biological Evaluation of Toxicity Caused by Combustion of Building Materials. *Fire Material.* **4**: 127-43.

Herpol, C., and Vandevelde, P. (1983) Effects of Furnace Direction on Toxicity Results in a Dynamic Moving Tubular Furnace Method. *J. Combust. Toxicol.* **9**: 165-186.

Hilado, J., Casey, C. J., Christensen, D. F., and Lipowitz, J. (1978) Toxicity of Pyrolysis Gases from Silicone Polymers. *J. Combust. Toxicol.* **5**: 130-140.

Hilado, C. J. (1976) Relative Toxicity of Pyrolysis Products of Some Foams and Fabrics. *J. Combust. Toxicol.* **3**: 32-60.

Hilado, C. J. (1976) Relative Toxicity of Pyrolysis Products of Polymeric Materials Rising Various Test Conditions and Ranking Systems. *J. Consum. Prod. Flamability.* **3**: 288-297.

Hilado, C. J. and Brauer, D. P. (1979) How Test Conditions and Criteria Affect Pyrolysis-Gas Toxicity Findings. *Mod. Plast.* **56**: 62-64.

Hilado, C. J., and Cumming, H. J. (1978) Relative Toxicity of Pyrolysis Gases From materials: Effects of Chemical Composition and Test Conditions. *Fire and Materials* **2**: 68-79.

Hilado, C. J. and Huttlinger, N. V. (1978) Toxicity of Pyrolysis Gases from Phenolic, Isocyanurate, and Polystyrene Rigid Foam Insulation. *J. Therm. Insul.* **2**: 40-7.

Hilado, C. J. and Schneider, J. E. (1979) Toxicity of Pyrolysis Gases from Polytetrafluoroethylene. *J. Combust. Toxicol.* **6**: 109-116.

Hilado, C. J. and Schneider, J. E. and Brauer, D.P. (1979) Toxicity of Pyrolysis Gases from Polypropylene. *J. Combust. Toxicol.* **6**: 109-116.

Hilado, C. J., and Solis, A. N. (1977) Some Possible Reference Materials for Fire Toxicity Tests. *J. Combust. Toxicol.* **4**: 163-172.

Hilado, C. J., Soriano, J. A., and Kosala, K. L. (1977) Effect of Heating Rate on Toxicity of Pyrolysis Gases from Some Synthetic Polymers. *J. Combust. Toxicol.* **4**: 533-555.

Hilado, C. J., Kosola, K. L., Solis, A. N., Kourtides, D. A., and Parker, J. A. (1977) *Toxicity of Pyrolysis Gases From Elastomers*. Washington, DC: National Aeronautics and Space Administration NASA TM-78-461.

Hinderer, R. K. (1984) A Comparative Review of the Combustion Toxicity of Polyvinyl Chloride. *J. Fire Sci.* **2**: 82-97.

Hoffman, H. T. and Sand, H. (1974) Further Investigations into the Relative Toxicity of Decomposition Products Given Off from Smoldering Plastics. *J. Fire Flamability Combust. Toxicol.* **Suppl 1**: 250-258.

Jellinek, H. H. G. and Takada, K. (1977) Toxic Gas Evolution from Polymers: Evolution of Hydrogen Cyanide from Polyurethanes. *J. Polym. Sci. Polym. Chem. Ed.* **15**: 2269-2288.

Johnston, P. K., Doyle, E. and Oryel, R. A. (1988) Acrylics: A Literature Review of Thermal Decomposition Products and Toxicity, *J.A.C.T.* **7**: 139-200.

Johnston, P. K., Doyle, E. and Oryel, R. A. (1988) Phenolics: A Literature Review of Thermal Decomposition Products and Toxicity. *J.A.C.T.* **7**: 201-220.

Kaplan, H. L., Grand A. F., Switzer, W. G., and Gad, S. C. (1984) Acute Inhalation Toxicity of the Smoke Produced by Five Halogenate Polymers. *J. Fire Sciences* **2**: 153-172.

Keller, J. G., Herrera, W. R. and Johnston, B. F. (1976) An Investigation of Potential Inhalation Toxicity of Smoke from Rigid Polyurethane Foams and Polyester Fabrics Containing Antiblaze 19 Flame Retardant Additive. *J. Combust. Toxicol.* **3**: 296-304.

Kimmerle, G. and Prager, F. K. (1980) The Relative Toxicity of Pyrolysis Products. Part I. Plastics and Man-made Fibers. *J. Combust. Toxicol.* **7**: 42-53.

Kishitani, K. and Nakamura, K. (1974) Toxicities of Combustion Products. *J. Fire Flam. Combust. Toxicol.* **Suppl 1**: 104-123.

Kishitani, K., and Nakamura, K. (1976). Study on Toxicities of Combustion Products of Building Materials at Initial Stage of Fire. **In**: *Proceedings of the Second Joint Meeting, U.S.-Japan Panel on Fire Research and Safety*. UJNR Tokyo, Oct. 19-22, pp.1-31.

Kourtides, D. A., Gilwee, W. J., and Hilado, C. J. (1978) Relative Toxicity of the Pyrolysis Products From Some Thermoplastic and Thermoset Polymers. *Polymer and Science* **18**: 674-676.

Levin, B. C., Fowell, A. T., Birky, M. M., Paabo, M., Stolte, A., and Malek, D. (1982). *Further Development of a Test Method for the Assessment of the Acute inhalation Toxicity of Combustion Products*. NBSIR 82-2532, U.S. Department of Commerce, National Bureau of Standards, National Engineering Laboratory, Center for Fire Research, Washington, D.C. (June 1982).

Liang, Y., Lu, P., Jiang, X., Chen, Z., Lu, Q., Ma, L., Znu, J, Zheng, H. and Fang, L. (1980) A Preliminary Study on Maximum Allowable Concentration of Pyrolysis Products of Poly(tetrafluorethylene) in Air Shang-hai Ti I.I. *Hsuch Yuan Pao* 7:276-82.

Liepins, R. and Pearce, E. M. (1976) Chemistry and Toxicity of Flame Retardants for Plastics. *Environ. Health Perspect.* **17**: 55-63.

Lieu, P. J., Magill, J. H. and Alaire, Y. C. (1979) Toxicity and Thermal Properties of Polyphosphazenes. SETEC - MME - 79-40 37p.

Lieu, P. J., Magill, J. H. and Alaire, Y. C. (1979) Toxicity of Degradation Products from Polyphasphazene Copolymers. Part IV SETEC - MME - 79-62 31p.

MacFarland, H. N. and Leong, K. J. (1962) Hazards from the Thermodecomposition of Plastics. *Arch. Environ. Health.* **4**: 591-597.

McGuire, P. S. and Annau, Z. (1980) Behavioral Effects of Exposure to the Combustion Products of Flexible Polyurethane Foam. *Neurobehav. Toxicol.* **2**: 355-361.

Michael, J. (1976) Toxicity of Pyrolysis and Combustion Products of Poly(vinyl Chloride). *Fire Mater.* **1**: 57-62.

Michael, J., Mitera, J., and Tardon, S., (1976) Toxicity of Thermal Degradation Products of Polyethylene and Polylpropylene. *Fire Mater.* **1**: 160-8.

Miller, D. P., Petrella, R.V., and Manca, A., (1976) Smoke and Toxic Gas Emission from Burning Unsaturated Polyester Resins. *Mod. Plast.* **53**: 97-103.

Mischutin, V. (1978) Non-Toxic Flame Retardants For Textiles. *J. Coated Fabrics* **7**: 308-318.

Nelson, G. L., Hixson, E.J. and Denine, E.P. (1978) Combustion Product Toxicity Studies of Engineering Plastics. **In**: *Proc. Int. Conf. Fire. Saf.* **3**: 123-139.

O'Mara, M. M. (1973) Analysis of the Combustion Products from Wood and Synthetic Polymers; Symposium: Products Combustion (Plastic) Building Materials 46-58.

O'Mara, M. M. (1974) A Comparison of Combustion Products Obtained from Various Synthetic Polymers. *J. Fire Flamability Combust. Toxicol.* **Suppl 1**: 141-156.

Paciorek, K. L., Kratzer, R.H. and Kaufman, J. (1973) Oxidative Thermal Degradation of Polytetrafluoroethylene. *J. Polym. Sci.* **11**: 1465-1473.

Parent, R. A., Dilley, J. V., Martin, S. B. and McKee, R. G. (1979) Acute Toxicity in Fischer Rats of Smoke from Non-flaming Combustion of Ten Flexible Polyurethane Foams. *J. Combust. Toxicol.* **6**: 185-197.

Paulson, D. R. and Moran, G. F. (1974) Analysis of Some Toxic Combustion Products of Low-Density Flexible Polyurethane-Ether Foams. *Environ. Sci. Technol.* **8**: 1111-1118.

Petajan, J. H., Voorhees, K. J., Packham, S. C., Baldwin, R. C., Einhorn, I. N., Grunnet, M. L., Dinger, B. G., and Birky, M. M. (1979) Extreme Toxicity from Combustion Products of a Fire-Retarded Polyurethane Foam. *Science* **187**: 742-4.

Pfaffli, P. and Zitting, A. (1978) Thermal Degradation Products of Homopolymer Polystyrene in Air. **In**: *Proceedings of the International Symposium on Styrene* 5-6.

Polakoff, P. L., Busch, K. A., Okawa, M. T. (1974) Urinary Fluoride Levels in Polytetrafluoroethylene Fabricators. *Amer. Ind. Hyg. Assoc. J.* **35**: 99-106.

Potts, W. J. and Lederer, T. S. (1977) A Method for Comparative Testing of Smoke Toxicity. *J. Comb. Toxicol.* **4**: 114-162.

Punderson, J. O. (1981) A Closer Look at Cause and Effect in Fire Fatalities - The Role of Toxic Fumes. *Fire and Mater.* **5**: 41-46.

Purohit, V. and Orzel, R. A. (1988) Polypropylene: A Literature Review of Thermal Decomposition Products and Toxicity. *J.A.C.T.* **7**: 221-240.

Robles, E. G. (1968) Thermal Decomposition Products of Cellophane EH2-M-68M-16 17pp.

Saito, Fumiharu (1970) Thermal-Decomposition Characteristics of Plastics-Mainly of PVC. *Enka Biniir To Porima* **7**: 6-17.

Saito, F. (1978). Factors accelerating the gas toxicity in a fire **In**: *Proceedings of the Joint Meetings. United States Japan panel on Fire Research and Safety.* UJNR, Washington, DC, March 13-18, 1978.

Sangha, G. K., Matijak, M. and Alarie, Y. C. (1981) Toxicologic evaluation of thermoplastic resins at and above processing temperatures. *J. Ind. Hyg. Assoc.* **42**: 481-485.

Scheel, L. D., Lane, W. C., and Coleman, W. E. (1968) Toxicity of Polytetrafluoroethylene Pyrolysis Products, including Carbonyl Fluoride and a Reaction Product, Silicon Tetrafluoride. *Am. Ind. Hyg. Assoc. J.* **29**: 41-8.

Seemann, J. and Woelcke, U. (1976) The Formation of Toxic Isocyanate Vapors from the Thermal Decomposition of Polyurethane Lacquers and Their Polyfunctional Hardeners. *Zentralbl Arbeitsmed, Arbeitsschutz Prophyl.* **26**: 2-9.

Sharrod, D. and Edser, M. H. (1976) Fire Hazard Characteristics of Some Latex and Emulsion Polymers Used in Textiles. *Text. Inst. Ind.* **14**: 79-82.

Smith, E. E. (1972) Fire Hazard Characteristics of Duct Materials. *ASHRAE* **14**: 29-35.

Smith, L. W., Gardner, R. J. and Kennedy, G. L. (1982) Short-Term Inhalation Toxicity of Perfluoroisobulylene. *Drug Chem. Toxicol.* **5**: 295-303.

Stark, G. W. V. (1969) Toxic Gases from PVC (Poly(vinyl Chloride)) in House-hold Fires. *Rubber Plast. Age* **50**: 283-4.

Sumitomo, B. (1974) Toxic Gases from the Combustion of Polyurethane Foams. *Purasuchikkusu* **25**: 11-22.

Tatem, P. A. and Williams, F. W. (1975) Development of Fire Performance Tests for Carbon-Impregnated Polyurethane Foams. *Ind. Eng. Chem.* **14**: 204-210.

Terrill, J. B., Montgomery, R. R., and Reinhardt, C. F. (1978) Toxic Gases from Fires. *Science* **200**: 1343-1347.

Tsuchiya, Y. and Sumi, K. (1967) Thermal Decomposition Products of Polyvinyl Chloride. *J. Appl. Chem.* **17**: 364-6.

Warity, R. S. and Kwon, B. K. (1968) Inhalation Toxicity of Pyrolysis Products of Poly(tetrafluoroethylene) Heated Below 500 Degrees Centigrade. *Am. Ind. Hyg. Assoc. J.* **29**: 19-26.

Williams, S. J., and Clarke, F. B. (1983) Combustion Product Toxicity: Dependence on the Mode of Product Generation. *Fire Mater.* **7**: 96-97.

Woolley, W. D., Ames, S. A., and Fordell, P. J. (1979) Chemical Aspects of Combustion Toxicology of Fires. *Fire Mater.* **3**: 110-120.

Wong, K. L., Stock, M. F., and Alarie, Y. C. (1983) Evaluation of the Pulmonary Toxicity of Plasticized Polyvinyl Chloride Thermal Decomposition Products in Guinea Pigs by Repeated CO_2 Challenges. *Tox. Appl. Pharmacol.* **70**: 236-248.

Yoshida, Y., Kono, K., Harada, A., Toyota, S., Watanabe, M., and Iwasaki, K. (1978) Toxicity of Pyrolysis Products of Thermal-Resistant Plastics including Polyamide and Polyester. *Nippon Zasshi* **33**: 450-8.

Young, W., Hilado, C. J., Kourtides, D. A., Parker, J. A., (1976) A Study of the Toxicology of Pyrolysis Gases from Synthetic Polymers. *J. Combust. Toxicol.* **3**: 157-165.

Zaborowski, L. M., Hill, J. W., and Wehlking, J. (1972) Polymer Pyrolysis. Hydrogen Cyanide from the Open Burning of Orion. *Environ. Lett.* **3**: 267-270.

Zitting, A., Pfaffli, P., and Vainio, H., (1978) Effects of Thermal Decomposition Products of Polystyrene on Microsomal Enzymes and Glutathione Levels in Mouse Liver. **In**: *Proceedings of the International symposium on Styrene: Occupational and Toxicological Aspects* pp 16-17.

Zitting, A. and Savolainen, H. (1980) Effects of Single and Repeated Exposures to Thermo-Oxidative Degradation Products of Poly (Acrylonitrile-Butadiene-Styrene) (ABS) on Rat Lung, Liver, Kidney, and Brain. *Arch. Toxicol.* **46**: 295-304.

Chapter 8

INFLUENCE OF FLAME RETARDANTS AND OTHER ADDITIVES ON THE TOXICITY OF COMBUSTION PRODUCTS

Shayne C. Gad

INTRODUCTION

Most of this volume concentrates on the toxicology of the thermal decomposition products of plastics considered as just the polymers themselves. However, it is extremely rare to encounter any "pure" plastics. Rather, almost all commercial plastics are "compounded" with one or (more often) more additives to improve their manufacturing and end-use commercial characteristics, and make them more attractive to customers and users. There are a great number of different varieties of additives whose basic toxicologic characteristics are reasonably well established; not so the characteristics of their combustion products.

The special case of flame retardants in polymers, in terms of their abilities to alter actual combustion toxicity (and the results of toxicity tests, which may very well be a separate function), is a complex, yet key, question. This is a group of plastic additives (the second largest volume additive, after plasticizers, which represent 6-7% of total plastic consumption according to Liepins and Pearce, 1976) which have been developed specifically to increase the safety of plastics in regards to fire - and yet have been the source of marked concern as regards health effects over the last ten years.

For example, tris-(2,3-dibromopropyl) phospate has now been well established as a mutagen and possible cancer hazard (Blum, 1977). And the reactive phosphorus flame retardant (0,0-diethyl-N,N bis-(2-hydroxyethyl)-aminomethyl phosphonate should be remembered as the source of much of the current regulatory effort in combustion toxicity. The Flamability Research Center in Utah discovered in 1975 (Petajan *et al.*, 1975) that this material, at levels of four and eight percent in rigid polyurethane foam, generated a thermolysis stream containing 4-ethyl-l-phospha-2,6,7-trioxabicyclo (2.2.2) octane-l-oxide, an extremely potent neurotoxic agent which induced grand mal-type seizures in animals within an hour of exposure. These results have since been reproduced by others, who also determined that another cyclic phosphonate (Antiblaze 19) did not produce these results (Keller *et al.*, 1976).

Conversely, there are studies in the literature which establish that other flame retardants decrease the toxicity of thermolysis streams in most cases. Hilado (1976) reported that the addition of either halogenated organophosphates or antimony troixide flame retardants to flexible polyurethane foams reproducibly decreased toxicity. Alarie and Anderson (1979) likewise examined the toxicity of several polymers with and without flame retardants and determined that their final index of toxicity, the LTC_{50}, increased (and therefore indicated decreased toxicity) in three out of four cases examined.

The key to the seeming puzzle of the influence of flame retardants on combustion toxicity is that there are several different factors acting. First, as O'Mara (1973) observed, additives (and particularly flame retardants) have marked effects on CO generation. Second, by definition flame retardants alter the temperature at which ignition occurs and the rate at which flames spread. Finally, in some cases the retardants themselves may be decomposed to generate materials of significant toxicity. This last factor is, of course, of greatest concern.

CLASSES OF ADDITIVES

There are at least nine major classes of additives for plastics. Though this chapter will focus primarily on one of these (flame retardants), the other eight also need to be considered at least to some extent.

Fillers

Fillers are materials which are frequently added to plastics to add bulk and to improve their strength, rigidity, sheer resistance, and dimensional stability (that is, ability to retain a shape). The resulting materials are frequently called "reinforced plastics" (Deanin, 1975; Brydson, 1966).

The major group of fillers (also called reinforcers) are the fibers, which generally are chemically inert. These include glass, asbestos (not much used anymore) and wood flour.

Many common inorganic mineral powders (such as calcium carbonate, silicates and glass) are also used as fillers, particularly in vinyls and polyesters.

Coupling agents (such as organosilanes) are sometimes used in small concentrations to ensure that polymers and fillers hold together and improve their water resistance.

Plasticizers

These are high molecular weight, high boiling point liquid monomers which are added to increase the flexibility and softness of the final product. They may be used in concentrations as high as 20 to 50% of the final composition (Brydson, 1966; Modern Plastics Encyclopedia, 1973). The most common plasticizers are the phthalates, which have been shown to cause cancer at high oral doses in rodents.

Organic phosphate esters are used, though only in lower concentrations. Some of these are delayed peripheral neuropathy inducing agents.

Also used are epoxidized esters (such as of soy bean oil and oleic esters), linear aliphatic esters (particularly for imparting low temperature flexibility), and polyesters.

Colorants

Plastics are generally colored by incorporating quantities of coloring agents into them during the production process.

Most of these colorants are common inorganic pigments such as titanium dioxide, iron oxides, and cadmium, chromium and molybdenum specialty pigments. Organic colors are also used.

Stabilizers

Many organic polymers are not generally stable enough for use in many modern processing steps (such as high temperature or pressure extrusion) or for long term use in many applications. Most plastics are improved in this regard by the addition of from 0.1 to 10% of stabilizers.

Polyvinyl chloride tends to shed HCl, leading to brittleness and rigidity. Several groups of thermal stabilizers serve to retard this loss. These include mixtures of metal phosphites or soaps (such as barium, cadmium and zinc), lead compounds (which also improve electrical isolating properties), and organotin compounds.

A second class of stabilizers are the antioxidants, which act to prevent or retard atmospheric oxidation reactions. Most common is butylated hydroxytoluene (BHT), which is also widely used in foods. Many other phenols and their sulfide dimers, dydrazides and triazoles are also used at concentrations of from 0.1 to 1.0%.

The atmospheric oxidation and degradation of most polymers is accelerated by the ultraviolet radiation in sunlight. Plastics that are to be used outdoors are therefore frequently stabilized against ultraviolet light. Most commonly used are such UV absorbers as orthohydroxyphenol benzotriazoles and orthohydroxybenzophenones, in concentrations from a few tenths of a percent to one percent. Salicyclic esters are also occasionally used because they are cheaper.

Other causes of the breakdown or reduced stability of plastics are mold, fungus and bacteria which can feed on compounds in the plastic (such as plasticizers). Copper quinolinate, quaternary ammonium compounds, and tributyltin oxide are commonly used as biocides.

Processing Aids

Many different materials are added to plastics in small quantities to improve their performance during processing. The first group of these are lubricating materials, used to improve the flow of liquid phase plastics. These include calcium, zinc and lead stearates, petroleum and polyethylene waxes, fatty esters and amides.

Flow control agents are used to cause plastics to evenly coat papers or foams to form evenly.

Curing Agents

These are used mainly for crossliking and otherwise finishing the polymerization process of such polymers as reenforced polyesters, silicone rubbers and other specialty elastomers, and polyethylene for use in wire and cable coating. These agents are generally peroxides such as benzoyl peroxide. A wide variety of peroxides are used in small amounts.

Antistatic Agents

In several applications (*e.g.* electrical insulators and carpet fibers) plastics can build up static charge. These charges are, at best, a nuisance and can also be dangerous. Most commonly used to avoid this are amines, quaternary ammonium compounds, polyoxyethylene glycol esters, and organic phosphates. The combustion toxicity of these has not been evaluated.

Stain Resistant Agents

For polymers used as fibers in household furnishings (such as carpets and cloth materials for sofas), it has become common practice in the last ten years to treat the fibers to make them resistant to staining and easy to clean. This can be done either by putting a finish on the fibers or by incorporating a stain resistant material into the polymers before the fibers are formed. Most commonly used are organolfuoride compounds, though the exact composition of these is generally held to be proprietary.

Classes of Flame/Fire Retardants

The largest and most important class of additives for consideration in combustion toxicity are the flame and fire retardants.

Basically, flame retardants function under heat to yield products that would be more difficult to ignite than the virgin plastics, or that do not propagate flame as readily. They function in one or more of the following ways (Frados *et al.*, 1981; Levek, 1978).

- By absorbing heat, thereby making sustained burning more difficult.
- By forming a nonflamable char or coating that insulates the substrate from the heat, excludes oxygen, and slows the rate of diffusion of volatile, flamable pyrolysis fragments from the substrate.
- By enhancing the decomposition of the substrate, thereby accelerating its melting at lower temperatures so that it drips or flows away from the flame front.
- By evolving products that stop or slow flame propagation.
- By forming free radicals that convert a polymer to less combustible products.
- By excluding oxygen from possible burning sites by coating resin particles.

Fire-retardant chemicals available commercially for plastics can be divided into two general classes, unreactive additives and reactive monomers or crosslinking agents. The unreactive additives are generally added to the polymer during processing, but do not react chemically with the other constituents of the composition. The reactive types, on the other hand, are generally chemically reacted with the polymer structure at some processing stage. The ideal fire retardant additive should be inexpensive, colorless, easily incorporated into the polymer composition, compatible, stable to heat and light, efficient in its fire-retardant properties, nonmigrating, and have no adverse effects on the physical properties of the polymer. The toxicology of the additive is also of increasing concern. Unfortunately, most presently available additives seldom meet all these requirements (Hendersinn, 1977).

Antimony Compounds

These are expensive and therefore used mostly in high value resins, particularly the high performance engineering thermoplastics, such as nylon, polybutylene teraphthalate and polyethylene terephthalate. These new polymeric grades still utilize antimony oxide as a synergist to achieve the desired flame resistance. multimetallic synergists which contain antimony and other metallic salts can also be used to give the same performance as antimony oxide at lower cost (Avento, 1981). Antimony oxide by itself is essentially useless as a fire retardant additive. Nevertheless, in combination with other materials, it is by far the most widely used antimony-containing flame-retardant additive. It is generally used with bromine- or chlorine-containing compounds. Combinations of antimony oxide and halogen compounds are primarily used as flame retardants in plastics such as ABS, polyethylene, polypropylene, polystyrene, polybutylene terephthlate (PBT) and unsaturated polyester. Depending on the severity of the flamability test specified and which plastic is being flame retarded, total use levels of such combinations range from 5-75 parts per 100 parts of plastic (Levek, 1978).

Antimony oxide is used in many formulations because it is so very effective when used in combination with a halogen. Concentrations of 30-40% by weight are required to give the desired properties. However, its use often elevates processing temperatures and causes loss of physical properties, such as tensile strength, percent elongation, and impact strength, when used in heavy loadings. On the other hand, antimony oxide can enhance light and oxidation stability, heat distortion and stiffness. It acts as a very strong white pigment, so colored fire-retardant polymers using this material usually are in pastel shades (Touval and Waddell, 1972).

Since antimony oxide is a dry white powder, some dusting occurs during handling, but its high density permits the dust to rapidly settle. Grades specially treated to reduce dusting are available. Particle size is an important consideration, since it can affect physical properties, color control in the finished product, and dispersibility or processibility.

Very recently, an announcement was made of the availability of colloidal antimony oxide (Sb_2O_5) for fire-retarding translucent reinforced plastic (unsaturated polyester) laminates (Whittaker and Kintz, 1982). The use of this material allows visual inspection of laminate quality during and after fabrication, and the retardant does not settle out of solution after being mixed into the resin.

chlorine compounds

chlorine compounds used as fire retardants include polychloroprene, PVC, chlorinated polyethylene, chlorosulfonated polyethylene, chlorinated paraffins, tris (chloroethyl) phosphate, tris (dichloropropyl) phosphate, tris (chloropropyl) phosphate, methyl pentachlorostearate, cycloaliphatic chlorine-containing flame retardants, and chlorendic anhydride (Frados, 1981; Levek, 1978). The cholorinated paraffins are the most widely used.

Bromine Compounds

Bromine compounds have also been used successfully. They are about twice as efficient as chlorine on a weight basis, so that significantly lower concentrations are needed. However, they are higher in cost than the chlorinated compounds and are generally less stable, thermally and under exposure to light (Hendersin, 1977). The bromine-containing additives are usually subdivided according to their structure. Those containing aromatic bromine are significantly more stable to heat and hydrolysis than the aliphatic type. Examples are decabromodiphenyl oxide (DBDPO) and tetrabromobisphenol A. The latter is used in both thermosets (epoxies and polyesters) and thermoplastics (styrene and polycarbonate). A pentabromodiphenyl oxide blend is now available for urethane foams and polyesters. Ferro Corporation has developed a new polymeric type of aromatic bromine flame retardant. Named Pyro-Check G8PB (Von Hassell, 1980), this compound is reported to be thermally stable and to have both nonflaming qualities and good toxicological properties, by virtue of its high molecular weight and polymeric nature. The new additive has an aromatic bromine content of 68% and can provide excellent levels of Underwriters Laboratories (UL) performance, depending on the use level and application. It can be combined with antimony oxide as a synergist to give UL acceptable performance. This compound is described as a benzene-ethenyl-tribromo-derivative homopolymer.

Another addition to the list of aromatic bromine compounds used is Firemaster 935, a product containing about 65% bromine. It is a polymeric product belonging to a class of polyphenylene oxides (Burleigh, 1979). This additive has been used very successfully with nylon 66 and PBT, both of which require high processing temperatures. Antimony oxide is usually used as a synergist with this compound. As with the Ferro product, flame spread (bloom) and toxicity are reported to be low.

Aliphatic bromine additives include hexabromocyclododecane, hexabromobutene, proprietary bicyclic oxides, and tris (dibromopropyl) phosphate. The latter compound has been one of the most widely used additives containing aliphatic bromine, but its toxicity is questionable and it's as a topical fire retardant for polyester fabric in childrens' sleepwear (Levek) is no longer allowed. Aliphatic bromine-type additives are used as flame retardants in plastic foam (polyurethane and polystyrene) and for polyester fibers. They are limited by their low thermal stability, however.

Ionic bromine-containing additives include ammonium bromide (used as a flame retardant in cellulosics) and phosphonium bromide (used to flame-retard thermoplastics), particularly in synergistic combination with other flame retardants such as the halogen compounds (Frados, 1981).

Phosphorus Compounds

Organophosphorus compounds are used to help flame-retard flexible vinyl systems. These materials plasticize, as well as contribute to the flame-retarding process. Modified polyphenylene oxide (PPO) systems also utilize organophosphates to achieve the balance of flame resistance and heat deflection required.

Halogenated organophosphorus compounds are used widely in flame retarding polyurethane foams (Avento, 1981). Some special grades recently made available include (i) an additive combining an intumescent compound (a phosphorus- and nitrogen-based additive) with ammonium polyphosphate, that has been suggested for use with polypropylene, polyethylene, polystyrene, and ethylene/propylene copolymers and (ii) a nonhalogenated phosphorus based retardant for use with polyolefins and other polymers.

A recent paper by Carnahan *et al.* reported investigations into the mechanism of phosphorus flame retardancy in engineering plastics. The use of elemental red phosphorus as a flame retardant (first used in 1965) has been reviewed by Peters and is rather surprising, since it will burn, under certain conditions, and is used in safety matches. Sometimes coadditives are necessary, and in a few instances red phosphorus has been reported to be ineffective with or without coadditives. usually it works only in a specified concentration range. It has been used successfully with thermoplastic polyesters, nylons, polyolefins, polyphenylene oxides, polyacetals, ABS resins, polycarbonates, styrenes, polymethylmethacrylate, and PVC. In some cases it acts synergistically with halogens.

Boron Compounds

Boron additives have also been found to have a synergistic effect with halogenated compounds, and are particularly attractive because they are lower in cost than antimony oxide (Hendersinn, 1977). Zinc borate is used as a partial replacement for antimony oxide in a number of applications, such as unsaturated polyester, polypropylene, and flexible PVC. Other boron compounds used are ammonium fluoroborate, boric acid and borax. Boron compounds are also useful in reducing or eliminating after-glow in halogen-containing compositions. Reports on the properties of Firebrake ZB, a special form of zinc borate ($2ZnO.3B_2O_3.35H_2O$) say that this compound retains its water of hydration at temperatures up to 500°F, which allows it to be used in polymers requiring high processing temperatures, such as nylon. The material has a low specific gravity (2.8), low toxicity, and a refractive index close to that of most polymers, allowing retention of translucency.

Firebrake ZB is synergistic with antimony oxide (Shen and Sprague, 1981, 1982). It costs less per pound than antimony oxide, is only half as dense, and is reported to reduce smoke up to 40% more than an equal volume of antimony oxide in certain halogenated polyesters and PVC.

Boron compounds act in the condensed phase by redirecting the decomposition process in favor of carbon formation rather than CO or CO_2. A second mechanism may involve the formation of a surface layer of protective char, which prevents the oxidation of carbon by limiting the accessible oxygen. In both cases a char is formed (Pearce *et al.*, 1975).

Aluminum Compounds

Alumina trihydrate (ATH), also known as hydrated alumina, has the chemical formula $Al(OH)_3$. Alcoa points out that its formula is frequently incorrectly written as $Al_2O_3.3H_2O$, which is really aluminum oxide trihydrate. It is unique among low cost fillers and extenders for plastics due to its high proportion (34.6%) of chemically bound water. This water of hydration is stable and unreactive at the processing temperature of many plastics. When heated to approximately 220°C or higher by exposure to a flame front, the hydrate decomposes into its constituents, with the absorption of a considerable amount of heat (Bonsignore, 1981).

Alumina trihydrate is an extremely important fire retardant for use in plastics. Its usage is close to 50% of all the additive fire retardants used for plastics (Anon, 1980). It is used primarily in unsaturated polyesters, where most applications can tolerate a high level of filler (Avento, 1981). Other applications include carpet backing for SBR latex, wire and cable formulations, phenolics, epoxies, and polyurethane foam. Polyester applications include shower and tub assemblies, plumbing fixtures, boat hulls and building panels. In most applications loadings are high (in excess of 40 parts per 100 parts of basic resin) (Levek, 1978). Although widely used in thermosets, it is also used in thermoplastics like polypropylene and ethylene/rubber compounds. ATH is a dry, light powder that functions by absorbing heat, by evolving steam to dilute the combustible gases being generated, and by providing a nonflamable char barrier between the heat source and the material. It also acts as a smoke suppressant.

Polymeric coatings have been applied to ATH to reduce dust and provide better adhesion to reinforced polyester. They also control the effect of the filler material on the gel and cure rates of the resin, and provide increased translucency to reinforced plastic products, with loadings up to 65% ATH by weight.

Although hydrated alumina can function as a fire retardant by virtue of its endothermic water release, at lower levels of loading physical effects are more important and finely divided anhydrous alumina actually appears more effective (Martin and Price, 1968; Weil, 1975).

Molybdenum Compounds

Molybdenum compounds, such as molybdenum trioxide, ammonium dimolybdate and ammonium octamolybdate, are useful in combustion flame retardants and smoke suppressants, especially in flexible PVC and reinforced polyester. Smoke emissions are reduced 25-50% in molybdenum compounds used in halogenated polyester and rigid and flexible PVC.

Reactive Fire Retardants

Reactive flame retardant systems contain functional groups allowing them to be incorporated directly into the polymer structure. The main advantage of this type of fire retardant is the permanence of the fire retardancy imparted. In most cases chemically reacting the fire retardant into the polymer essentially eliminates long-term migration of the fire retardant. However, these materials may be more expensive in the long run because of the necessity of reacting the material with the polymer. In some cases the processing techniques are more difficult (Hendersinn, 1977).

Reactive flame retardants are primarily used in unsaturated polyesters, epoxy resins and polyurethane foams. Two of the most popular reactive retardants are tetrabromobisphenol A and dibromomoneopentyl and tetrabromophthalic glycol. Others include chlorendic acid anhydride, tetrabormophthalic and tetrachlorophthalic anhydride, and diallyl chlorendate. Other reactive flame retardants vary significantly in functionality and can be useful in many polymer reactions and applications. Reactive polyols which contain halogen groups, phosphorus, or both, are offered by a number of suppliers for flame retardant urethane foam applications. These materials can be used alone, or with other flame retardants as synergists. Although polymeric systems reacted with a flame retardant monomer may appear to be more costly initially, in the long run they may be found to be less expensive than the additive types.

Summary

For flame retardant resins using additive systems several negative factors besides cost increases and physical property degradation must be considered. Halide acids may corrode the processing equipment used. Ultraviolet resistance is sometimes reduced by flame retardant additives. Silicone mold release is highly flamable, and since some of this material remains on the surface of the plastic product, it temporarily reduces the expected fire retardance of a resin. This effect dissipates in 1-2 months. Also, several of the commonly used colorants adversely and permanently change the flame retardance properties. For this reason UL requires that colored resins be supplied in several colors before approval is granted (Zimmermann, 1979).

Hayes and Lesniewski suggest that, in some cases, it is wise and practical to formulate resins with retardants to meet the slightly less rigid UL 94 V-2 requirements, rather than use the V-0 or V-1 requirements and have the properties drop unacceptably. In certain cases, these workers point out, UL requires V-2 plastics to provide substantially greater resistance to ignition by electrical faults, such as hotwire ignition and arc ignition, than it requires for a V-0 or V-1 plastic for the same application. Further, many V-2 compounds are substantially superior because they do not form a conducting char (which often occurs with V-0 plastics) caused by an electric arc (Hayes and Lesniewski, 1981).

TOXICITY OF FLAME RETARDANT CLASSES

There is data on the combustion toxicity of five major classes of flame retardants used in polymers, though only in the case of the phosphorus containing compounds is this data substantial. Acute toxicity date of prototypical compounds is summarized in Table 8-1. Combustion toxicity data is presented below.

Bromine Containing

The halogen (bromine and chlorine) containing flame retardants display both similarities and differences. Some of these relationships are obvious. Miller *et al.* (1976), for example, reported that, as more bromine was added to a resin, a greater amount of HBr was generated, therefore increasing the toxicity of a polymer when present in large amounts.

Benbow and Cullis (1975) investigated decabromobiphenyl and decabromobipheryl ether and found that both volatilized unchanged if no flame was present. When a flame was present, however (as would be the case in most "real world" situations) hydrogen bromide was generated. Not

Table 8-1. Acute Toxicity Data on Flame Retardants

Aluminum Hydroxide (trihydrate)	Rat oral LD_{50} > 20g/kg
Antimony Trioxide	Rat oral LD_{50} > 20g/kg
Chlorendic Anhydride	Rat oral LD_{50} = 1.77g/kg
Decabromodiplenyl Oxide	Not an eye or skin irritant
Hexabromobenzene	Mouse oral LD_{50} > 7g/kg
Molybdenum Trioxide	Rat oral LD_{50} = 125mg/kg
	Rat oral LC_{75} = 431mg/m^3
TRIS-(2,3-dibromoprypyl) Phosphate	Rat oral LD_{50} = 1.01g/kg
	Mouse oral LD_{50} = 1.149g/kg
	Severe skin irritant
	Moderate eye irritant
Zinc Borate	Mouse oral LD_{50} > 2.5g/kg

detected under flaming conditions was the formation of lower phenyl oxides which would be much more toxic.

Mischutin (1976) evaluated decarbromodiphenyl oxide as a flame retardant for cotton/polyester blends and found that it did not alter their combustion toxicity.

Overall, it appears that the bromine-containing flame retardantscurrently in use represent a minimal hazard.

Chlorine Containing

Bromine containing systems are preferred over chlorine-containing systems for a number of reasons. As a result there has been less combustion toxicity testing performed on chlorine-containing systems. In the single complete study reported in the literature, Miller *et al.* (1976) found that the chlorine containing systems generated much higher levels of HCl than the analogous bromine containing systems generate of HBr. Additionally, they found that small amounts of HBr were also generated by the chlorine containing system - a not uncommon contamination finding with halogen compounds, caused by incomplete separation of the halogens in starting materials. Based on this, they predicted bromine systems are less hazardous than the chlorine systems.

Phosphorus Containing

Because of the TRIS mutagenicity findings, this group of flame retardants has attracted a greater amount of interest than have the other groups. The published studies can be readily classified as belonging to one of two groups - those concerned with mutagenic/ carcinogenic/teratologic effects and those

Table 8-2. Long Term Effects of Phosphorus Containing Flame Retardants

Reference	Material	Test/Endpoint	Results
Connor *et al.*, 1980	Tetrakis (Hydroxy-methyl) Phosphonium Sulfate	Ames/Mutagenicity	Negative
		In vivo Ames/Mutagenicity	Negative
		Micronuclues/Mutagenicity	Negative
		Metaphase Analysis/ Mutagenicity	Negative
Salamone and Katz, 1981	Tris (2,3-dibromopropyl) Phosphate	Micronucleus/Mutagenicity	Positive
		Ames/Mutagenicity	Positive
		Sperm head/Mutagenicity	Positive
Seabaugh *et al.*,	Tris (2,3-dibromopropyl) Phosphate	Rat teratology	Negative
Van Duren *et al.*, 1978	Tris (2,3-dibromopropyl)	Mouse dermal/ Carcinogenicity	Positive
	Tetrakis	Mouse dermal/Carcinogenicity	Negative

concerned with acute toxicity. The first of these groups is summarized in Table 8-2.

The results of the acute combustion toxicity studies also appear to vary considerably between researchers and compounds. miller *et al.* (1976) found little effect on the toxicity of combustion products generated by polyurethanes in the presence or absence of an unspecified phosphorus-containing flame retardant. Hilado *et al.* (1978) found that tetrakis (2-chlorethyl) ethylene diphosphate in flexible polyruethane foams and N,N-bis-(2-hydroxyethyl) aminomethyl phosphonate in rigid polyurethane foams caused no significant alterations in the combustion product toxicities as measured in the USF system. Kimmerle (1976) however, found that trimethylolpropane markedly increased the toxicity of a rigid polyurethane foam. Finally, one should keep in mind 0,0-diethyl-N,N-bis(2-hydroxyethyl)-aminomethyl-phosphonate, the potent neurotoxic-producing material which started the current wave of interest in the field (Petajan *et al.*, 1975). Napier and Wong (1972) established that wide varieties of organophosphates were generated by phosphorus-containing flame retardants over a wide range of thermal decomposition conditions.

Alumina Hydrates

Miller *et al.* (1976) found that alumina hydrate markedly decreased the generation of smoke and free hydrogen halide (HX). It is suspected, however, that the HX generated may be adsorbed onto the smoke particles that are generated.

In a previously unpublished study, the author evaluated a shredded cellulose insulation with and without alumina hydrate present. Done in the CMIR system, it was found that in the flaming mode the LC_{50} of the material was increased from 16.4 mg/L without (alumina hydrate) to 29.8 mg/L with. In the nonflaming mode the increase was 14.7 mg/L (without) to 23.4 mg/L (with). The data available and the chemistry of the material indicate that this semi-durable flame retardant represents no significant hazard.

Antimony Trioxide

No direct work has been published on the effects upon combustion toxicity of this material. In theory, it acts by forming a mist of antimony oxide upon volatilization, which then acts to impede flaming (Beabow and Cullin, 1975). It has been hypothesized that this would serve to increase the toxicity of combustion products by leading to the production of highly toxic antimony halides. There is currently no data to support this contention.

Green (1976) found that the presence of flame retardants increased the formation of carbon in the pyrolysis residue and decreased the yield of volatile aromatic products, which would argue for a reduction in actual smoke production.

It should be kept in mind that smoke consists of both a fine liquid and a solid aerosol component. Its presence is normally measured by light absorption, which is expressed as optical density. The optical density is proportional to the square of the average particle radius, which means the measurement is very sensitive to larger sized particles and droplets. But is does not give any true measure of either particle/droplet size or mass concentration. None of the current combustion toxicity methods measure smoke at all, but rather limit themselves to determining biological effects and the quantitative and qualitative aspects of the gas phase. If the biological effects are to be truly characterized this will need to be changed.

REFERENCES

Alarie, Y. C. and Anderson, R. C. (1979) Toxicologic and Acute Lethal Hazard Evaluation of Thermal Decomposition Products of Synthetic and Natural Polymers. *Toxicol. Appl. Pharmacol.* **51**: 341-362.

Anonymous. Flame Retardants Offer Something Extra. *Modern Plastics* **57**:L58-59 (February, 1980).

Avento, J. M. (1981) Flame Retardants. *Modern Plastics Encyclopedia*. **58(19A)**: 187-190.

Benbow, A. W. and Cullis, C. F. (1975) The Formation of Toxic Products During the Combustion of Halogen-Containing Polymers. *Eur. Polym. J.* **11**: 723-727.

Blum, A. (1975) Flame Resistant Additives as Possible Cancer Hazards. *Science* **195**: 17-23.

Bonsignore, P. V. (1981) Alumina Trihydrate as a Flame Retardant for Polyurethane Foams. *Journal of Cellular Plastics* **17**: 220-225.

Brydson, J. A. (1966) *Plastics Materials*. Van Nostrand. Princeton.

Burleigh, P. H., Nametz, R. C., Moore, P. O. and Joy, T. A. (1979) A New Flame Retardant for Engineering Thermoplastics. *Proceedings of the 4th International Conference on Fire Safety*. San Francisco, CA, Vol. 4, pp. 279-293.

Carnahan, J., Haaf, W., Nelson, G., Lee, G., Abolins, V. and Shank, P. (1979) Investigations into the Mechanism for Phosphorus Flame Retardancy in Engineering Plastics. *Proceedings of the 4th International Conference on Fire Safety*. San Francisco, CA, Vol. 4, pp. 312-322.

Connor, T. H., Meyne, J. and Legator, M. S. (1980) The Mutagenic Evaluation of Tetrakis (Hydroxymethyl) Phosphonium Sulfate Using A Combined Testing Protocol Approach. *J. Environ. Path. Toxicol.***4: 145-158.**

Deanin, R. D. (1975). Additives in Plastics. *Env. Hlth. Perspect.* **11**: 35-39.

Frados, Joel (Ed.). *Flame Retardants*. Plastics Compounding Redbook, 1981-82, pp. 39-40.

Green, R. J. S., Hume, V., and Kumar, S. (1976) Burning Plastics: mechanisms of Smoke Formation. *Fire and Mater.* **1**: 36-41.

Hayes, W. K. and Lesniewski, J. M. (1981) How to Meet Flame Retardancy Specs and Still Make a Profit. *Plastics Engineering* **37**: 25-31.

Hendersinn, R. (1977) Fire Retardancy (Survey). **In**: *Encyclopedia of Polymer Science and Technology* (N. B. Bikales, Ed.), Supplement, Vol. 2, pp. 270-339. Wiley-interscience, New York.

Hilado, C. J. (1976) Relative Toxicity of Pyrolysis Products of Some Foams and Fabrics. *J. Combus. Toxicol.* **3**: 32-60.

Hilado, C. J., Schneider, J. E. and Murphy, R. M. (1978) Effect of Fire Retardant on Relative Toxicity of Pyrolysis Gases from Polyurethane Foams. *J. Fire Retard. Chem.* **5**: 83-85.

Keller, J. G., Herrera, W. R. and Johnston, B. F. (1976) An investigation of Potential Inhalation Toxicity of Smoke From Rigid Polyurethane Foams and Polyester Fabrics Containing Antiblaze 19 Flame Retardant Additive. *J. Combust. Toxicol.* **3**: 296-304.

Kimmerle, G. (1976) Toxicity of Combustion Products with Particular Reference to Polyurethane. *Ann. Occup. Hyg.* **19**: 269-273.

Levek, R. P. (1978) Flame Retardant Additives for Plastics, **In**: *Additives for Plastics, State of the Art*. Vol. 1 (Raymon B. Seymour, Ed.) pp. 259-275. Academic Press, New York.

Liepins, R. and Pearce, E. M. (1976) Chemistry and Toxicity of Flame Retardants for Plastics. *Environ. Health Perspect.* **17**: 55-63.

Martin, F. J. and Price, K. R. (1968) Flamability of Epoxy Resins. *Journal of Applied Polymer Science* **12**: 143-158.

Miller, D. P., Petrella, R. V. and Manca, A. (1976) Smoke and Toxic Gas Emission From Burning Unsaturated Polyester Resins. *Mod. Plast.* **53**: 97-103.

Mischutin, V. (1978) Non-Toxic Flame Retardants for Textiles. *J. Coated Fabrics* **7**: 308-318.

Modern Plastics Encyclopedia (1973), Vol. 50(10A). McGraw-Hill. New York, p. 189.

Napier, D. H. and Wong, T. W. (1972) Toxic Products From the Combustion and Pyrolysis of Polyurethane Foams. *Br. Polym. J.* **4**: 45-52.

O'Mara, M. M. (1973) Analysis of the Combustion Products From Wood and Synthetic Polymers. Symposium: Products Combustion (plastic). *Building Materials* 46-58.

Pearce, E. M., Shalaby, S. W. and Barker, R. H. (1975) Retardancy of Combustion of Polyamides. **In**: *Flame Retardant Polymeric Materials* (Vol. 1), (M. Lewin, S. M. Atlas and E. M. Pearce, Eds.). Plenum Press. New York.

Petajan, J. H., Voorhees, K. J., Packham, S. C., Baldwin, R. C., Einhorn, I. N., Grunnet, M. L., Dinger, B. G. and Birky, M. M. (1975) Extreme Toxicity From Combustions Products of a Fire Retarded Polyurethane Foam. *Science* **187**: 742-4.

Peters, E. N. (1979) Red Phosphorus as a Flame Retardant. **In**: *Flame Retardancy of Polymeric materials*, Vol. 5, pp. 113-176 (William C. Kuryla and Anthony J. Papa, Eds.). Marcel Dekker, Inc., New York.

Salamone, M. F. and Katz, M. (1981) Mutagenicity of Tris (2,3-dibro mopropyl) Phosphate in Mammalian Gonad and Bone Marrow Tissue. *JNCI* **66**: 691-5.

Seabaugh, V. M., Collins, T. F., Hoheisel, C. A., Bierbower, G. W. and McLaughlin, J. (1981) Rat Tertology Study of Orally Administered Tris-(2,3-Dibromopropyl) Phosphate. *Food Cosmet. Toxicol.* **19**: 67-72.

Shen, K. K. and Sprague, R. W. (1982) Zinc Borate as a Flame Retardant and Smoke Suppressant in Epoxy Systems. **In**: *37th Annual Technical Conference*. Reinforced Plast /Composites Institute, Washington, DC, Paper 12-F, pp. 5.

Shen, K. K. and Sprague, R. W. (1982) A Unique Form of Zinc Borate for Flame Retarding Reinforced Polyester. *Plastics Compounding* **5**: 67-73.

Touval, I. and Waddell, H. H. (1972) What You Should Know About Flame Retardants, **In**: *Advances in Fire Retardants*, Part 1 (Vol. 2 in Fire Retardancy Series), pp. 124-134 (V. M. Bhatnagar, Ed.). Technomic Publishing Company. Westport, CT.

Van Duuren, B. L., Loewengart, G., Seidman, I., Smith, A. C. and Melchionne, S. (1978) Mouse Skin Carcinogenicity Tests of the Flame Retardants Tris (2,3-Dibromopropyl) Phosphate, Tetrakis (Hydroxymethyl) Phosphonium Chloride, and Polyvinyl Bromide. *Cancer Res.* **38**: 3236-3240.

Von Hassell, A. (1980) Flame Retardants: Signs of a "Thaw" in R and D. *Plastics Technology* **26**: 71-74.

Weil, E. D. (1975) Additivity, Synergism, and Antagonism. **In**: *Flame Retardancy of Polymeric Materials*, Vol. 3 (W. C. Kuryla and A. J. Papa, Eds.). Marcel Dekker, Inc. New York.

Whittaker, W. A. III and Kintz, D. P. (1982) Translucent Fire Retardant RP Laminates Utilize Colloidal Antimony Oxide. **In**: *37th Annual Conference, Reinforced Plastics/composites Institute*. Washington, DC, paper 17-C, pp. 5.

Zimmermann, A. (1979). Meeting Flame Retardancy Requirements. *Plastics Design Forum* **4**: 80-83.

Chapter 9

CURRENT STATUS, ISSUES AND DIRECTIONS IN COMBUSTION TOXICOLOGY

CURRENT STATUS

H. Kaplan

In the preface to this volume, the statement was made that combustion toxicology has regressed from a developing science to the technology of the median lethal concentration. This does not mean that all activity in the field, except determining LC_{50} values, have come to a standstill. On the contrary, in recent years testing, research and technology development efforts have probably increased, particularly in the development of combustion devices, standard test methods and models for hazard analysis and risk assessment. It is the intent of this section to examine current activities in combustion toxicology, and to evaluate what these activities are likely to accomplish.

New York State has legislated combustion toxicity testing, possibly the only activity not under the direction of the plastics industry. The legislation requires the submission of LC_{50} values (determined by the University of Pittsburgh test method) and other information, mostly unrelated to toxicity, for certain building and finishing materials. This costly three-year program, which is nearing completion, has resulted in the submission of hundreds, perhaps thousands, of LC_{50} values to the New York State Fire Marshall's Office. The voluminous data, which are required only for filing purposes and not for the regulation of materials, may necessitate construction of a storage facility. This unfortunate program has resulted in needless cost to the consumer, without likelihood of preventing even a single smoke inhalation fatality. Even worse, the use of these LC_{50} values by decorators, designers or architects could lead to the selection of materials that are actually more hazardous than alternatives.

The American Society of Testing Methods (ASTM) has been considering the University of Pittsburg test method and others as a standard ASTM test method for a number of years. With very few exceptions, the task groups responsible for various aspects of smoke toxicity have been dominated by chemists, engineers, marketing people and other non-toxicologists employed or funded by the industrial sector. It is not surprising that the organization had so much difficulty in obtaining concurrence even on a definition of smoke toxicity. In view of the large membership from the plastics industry in ASTM, there is little chance that the University of Pittsburgh method will be adopted as an ASTM test method. In fact, it is highly unlikely that the current committees and interested parties will permit the adoption of any test method unless the method shows insignificant differences in the toxicities of smokes generated by most materials.

Another test method for "measuring the combustion toxic hazard" of materials is presently under development by the National Institute of Building Sciences (NIBS), which is at least partially funded by industry. Typically, the working group responsible for guiding and monitoring the development of the method has had minimal representation by trained toxicologists and is chaired by a non-toxicologist. The method, announced by the NIBS as a new concept in 1988, is an attempt to measure the combustion toxic hazard of a material by integrating its five performance parameters and its smoke toxicity. The procedure has an improved radiant heat combustion device, uses lethality as the index of toxicity, and, unless changed recently, relates lethality to irradiation time. The method may be adequate for measuring fire performance parameters of a material but, with its reliance in lethality of rodents, will not provide an accurate measure of potential toxic hazard. For some reason, perhaps training in the physical sciences, the perennial members of toxicity or toxic hazard test method committees, working groups or task groups, equate lethality with toxicity or toxic hazard and are convinced that toxicity can be adequately described simply by the number of dead rodents. Based on experience with other radiant heat combustion devices, the NIBS method will probably understate the toxic hazard of irritant-generating materials relative to that of materials which produce CO and/or HCN. If adopted by ASTM or regulatory building code organizations, the method will have little impact on smoke inhalation fatalities and may actually result in the selection of more toxicologically hazardous materials.

Most of the current activities in combustion toxicology and related areas, such as development of combustion devices and hazard analysis/risk assessment, continue to be conducted at the Center for Fire Research (CFR) of the National Institute of Science and Technology (NIST), formerly called the National Bureau of Standards (NBS). One major effort of the CRF has been the n-gas model, which was proclaimed as a new approach to evaluating the toxicity or toxic hazard of smoke several years ago. This simplistic additivity "model" is essentially the additive formula for mixtures of the American Conference of Governmental Industrial Hygienists (ACGIH)

$$\frac{C_1}{T_1}+\frac{C_2}{T_2}+\frac{C_3}{T_3}+\ldots=1\,,$$

except that the threshold limit value (T) is replaced with the LC_{50} value of each component gas in the n-gas model. The significant difference is that the ACGIH recommends application of its formula only when the components in a mixture have similar toxicological effects, whereas the n-gas model attempts to add "apples and oranges". The model may be functional for certain concentrations of CO and HCN in a mixture, the additivity of which was scientifically demonstrated approximately 15 years ago, but its application in attempting to sum the lethal effects of asphyxiant and irritant gases is inappropriate. The latter is apparent from work reported by the CFR, in which the n-gas model failed to predict lethalities for approximately 30 percent of the materials tested.

Another major activity of the CRF is the development of hazard and risk assessment models. These models attempt to utilize the fire performance characteristics of materials, such as ignitability, flame spread and rate of heat release, as well as the distribution and transport of heat and smoke and the toxicity of smoke to assess the hazards and risks of materials and building configurations. Large-scale fire tests are conducted to provide input to these models for validation purposes. Funding for some of these activities is provided by the National Fire Protection Research Association (NFPRA). Very little information is available on the progress of these hazard analysis and risk assessment models. Although modeling of fire performance parameters and physical conditions in a fire appears possible, rodent lethality data will not provide an adequate assessment of the toxic hazards of fires to man. It is doubtful that these efforts will contribute significantly to improved fire safety, except, perhaps, for assisting in the selection of materials with better fire performance parameters.

It is truly unfortunate that an agency of a respected government organization, the NIST, with the responsibility for reducing fire fatalities has not shown leadership and made meaningful scientific contributions to this field. Instead, beginning with its five or six-year program in the 1970s that produced a combustion device, a plastic exposure chamber and an acute toxicity protocol, it has promulgated scientifically inaccurate statements and concepts, rediscovered and became infatuated with the LC_{50}, assembled countless symposia, meetings and committees, and funded grants described by peer reviewers as being of questionable value at best.

Another participant in the combustion toxicity area is the International Standards organization (ISO). In keeping with custom, the technical committee responsible for combustion toxicity is chaired by a non-toxicologist from U.S. industry and, in contrast to the European countries, participants from the U.S. are not trained toxicologists. This should not be surprising because funding for the U.S. representatives to ISO, with the exception of the NBS delegate, has been provided by industry. The activities of the working groups have been limited to writing technical reports on various aspects of combustion toxicology, including biological assay methods, analytical methods, and modeling and hazard analysis. The reports, which are essentially literature reviews, take several years to write and obtain the concurrence of members, and are generally obsolete by the time they are finalized.

It should be obvious to the reader that most, if not all, current activities in combustion toxicology, except for the New York State testing program, have at least the potential and appearance of being unduly influenced by the plastics industry. It also is obvious that input into this field from trained and qualified toxicologists has been kept, apparently intentionally, to a minimum. As long as these conditions persist, the potential combustion toxicity of materials to man will not be accurately assessed by any of the toxicity test methods. Nor will any of the models for hazard analysis and risk assessment adequately predict the toxic hazard of building materials and furnishings in fires. These deficiencies, however, may be irrelevant because, in the opinion of some, the conclusion of hazard analysis and risk assessment studies has already been formed. That conclusion is that carbon monoxide is produced by all materials in sufficient quantities to cause most smoke inhalation fatalities and that differences in the toxicity of smoke generated by materials make little difference in the resultant hazard of a fire. With this conclusion, the problem of smoke toxicity to the plastics industry can finally be laid to rest.

ISSUES AND DIRECTIONS (S.C. Gad)

As should be apparent by this point, the discipline of combustion toxicology has advanced to the point where it is on the verge of a transformation. To this point, as has been the case with much of toxicology in general, the field has progressed in the manner of a descriptive science - data which acts to describe what happens under a variety of situations has been collected, but no attempt has been made to explain why things happen. This second step, the development of models which explain why things happen and which therefore serve to predict as yet unobserved happenings, is the mechanistic phase of a science. Combustion toxicology is apparently on the verge of its mechanistic phase and has stalled at this point. This last brief chapter aims to point out the questions to be answered in order to move the field to the mechanistic phase, and the more obvious seeming discontinuities in the current practice of the field.

The questions to be addressed are of an objective or an operational nature. The questions of objective calls for consideration of what is the question to be answered by the test system employed. To date, those that are being required by one or another government organization are justified on the grounds that they serve to detect materials that represent unacceptable hazards. This is the classical operational requirement for a screen, and such proposed test systems should be carefully considered in terms of the functional characteristics of screens in toxicology (Gad, 1988).

The first operational question which must be addressed is that of the relationship between the large scale (or "real") fire and the various laboratory models which are currently being employed. Measurements of what goes on in the different regions of a real fire and studies of how particular materials behave in a full scale situation have been made. But only a very few attempts have been made to study the biological effects of large scale fire atmospheres, and these have not been published. Some of the laboratory scale models attempt to address this indirectly by incorporating evaluations of several diverse thermolysis conditions (flaming, nonflaming and 440°C for example) into the evaluations. But a true validation of any of the proposed test systems will require a series of parallel experiments (large scale and laboratory scale) which directly compare the results of defined loads of test materials. Only this will provide a verification of where the results of the laboratory tests are valid predictors and where they are not. The mathematical modeling of fires, which has been a primary objective of much of the large scale fire testing to date, has (unfortunately) progressed only to the stage of accounting for very simple fire situations. The modeled situations are even less real than the laboratory scale tests themselves. Likewise, the current mathematical models of the toxicity of complex combustion atmospheres are extremely limited in both utility and accuracy beyond cases of simple mixtures of a few well studied gases.

Three basic questions which must be answered before the field can be advanced to the mechanistic phase are the relationships between toxicity and variations in (1) available oxygen during decomposition, (2) alterations in heating rates, and (3) the generation and characteristics of aerosols in combustion atmospheres. Some work has been done on both of these questions but neither has been definitively addressed.

Originally, it was believed that the most dangerous thermolysis state in terms of the production of toxic products was the smouldering (or limited air available) condition. Very little work has directly addressed this point. Hilado and Brauer (1979) did present some evaluations of the effects of variations in air flow through the USF system, which does not quite address the problem. Other work already presented in this volume supports the view that limited oxygen in a thermolysis environment does lead to increased production of CO as opposed to CO_2. But a complete evaluation, particularly in the test systems now being used to regulate, is required.

The question of the influence of heating rates has been more widely recognized and addressed. Cornish *et al.* (1975) did evaluate the differences in toxicity of products of decomposition when the samples were slowly heated at 5°C/minute as opposed to a rapid heating and did find major differences in toxicity. Hilado *et al.* (1979) likewise studied LC_{50} values of 12 materials as they were affected by alterations in the final sample temperature and in heating rate. They believed that simple thermodynamics explained the wide variations that were observed.

Using the CMIR system, a series of studies (Gad and Smith, 1984) was conducted, some years ago, on the "ramping" effect - that is, the alterations in results produced by different heating rates. The equipment utilized was able to reproducibly control sample heating rates over a very wide range. A portion of these results are presented in Table 9-1. Not shown in the table is that measurable levels (0.5 - 5.3 ppm) of nitrogen oxides were seen, but only in the 50°C/minute runs. A number of patterns are clear in the peak levels of the different materials generated and the biological results (death and incapacitation) achieved. Further investigation would seem to be warranted.

Additionally, there is a major area where data, though of obvious significance, are not being collected by any of the current test systems. This is the issue of smokes and aerosols - what levels of which types are being generated, how do combustion conditions effect their generation, and what are their biological effects?

It was recognized early on (Zapp, 1951; Zikria *et al.*, 1972) that the smoke particles generated during thermolysis were of significance not just because of their effects, but also because they can serve as carriers to promote deep lung penetration of other materials which are adsorbed to their surfaces.

Hilado (1970), in his earliest published work on combustion toxicity, reported evaluations of the effects of physical factors, flame retardants, and combustion conditions on smoke evolution. Among the findings were that decreasing the density of cellular plastics decreases total smoke production, but also causes a more rapid generation of smoke. Some flame retardants were reported to increase smoke generation by causing larger particles to become entrained in the gas phase. Using the NBS smoke chamber and the associated optical smoke density method, he found that (in terms of smoke production)

TABLE 9-1. WOOD HEATING RATE STUDIES

Sample Heating Rate(1)	Fuel Load (2)	Normal Toxicant Concentrations (3)	Mortality results	Incapacitation Results	Time To First Incapacitation	Maximum CO_2 (4)	Maximum CO (%)	Maximum HC (ppm)
30 °	1 gm	10.5 mg/ml	0/6	0/6	N/A	*	*	*
	2 gm	19.4 mg/ml	0/6	3/6	26' - 54'	3.6	0.25	660
	3 gm	33.0 mg/ml	2/6	6/6	*	*	*	*
	4 gm	44.0 mg/ml	4/6	6/6	24' - 38'	2.3	0.55	43
				LC_{50} = 38.5 mg/ml				
				IC_{50} = 19.9 mg/ml				
20°	2 gm	18.6 mg/ml	0/6	0/6	N/A	3.1	0.25	240
	3 gm	28.1 mg/ml	0/6	0/6	N/A	*	*	*
	3 gm	41.3 mg/ml	0/6	6/6	24' - 38'	1.8	0.6	60
	4 gm	44.4 mg/ml	4/6	6/6	27' - 18'	3.1	0	580
				LC_{50} = 43.0 mg/ml				
				IC_{50} = 33.5 mg/ml				

Notes:
(1) °C per minute sample temperature was increased from 25°C to autoignition
(2) Weight of sample loaded into system
(3) Total weight loss of sample divided by system volume
(4) All gas levels are peak values measured - not integrated totals
* = Values not measured
HC - Hydrocarbons

plastics can be divided into two groups. The first are those which produce more smoke under nonflaming conditions: polyethylene, polypropylene, and styrene - acrylonitriles. The second are those which produce more smoke under flaming conditions, and included PVC, polystyrene, ABS, polysulfone, polycarbonate and polyesters.

Green (1976) found that the presence of flame retardants increases the formation of carbon in the pyrolysis residue and decreases the yield of volatile aromatic products, which would argue for a reduction in actual smoke production.

It should be kept in mind that smoke consists of both a fine liquid and a particulate aerosol component. Its presence is normally measured by light absorption, which is expressed as optical density. The optical density is proportional to the square of the average particle radius, which means the measurement is very sensitive to larger particles and droplets. But it does not give any true measure of either particle/droplet size or mass concentration. None of the current combustion toxicity testing methods measure smoke at all, but rather limit themselves to determining biological effects and the quantitative and qualitative aspects of the gas phase. If the biological effects are to be truly identified and characterized, this will need to be changed.

In previously unreported work, incapacitation time data obtained using the CMIR polymer thermolysis procedures were analyzed to see if it was of a normal nature. This was done by first transforming the members of each individual group (the 1 to 6 individual animal times obtained during a single exposure trial being treated as a group) to Z scores, then pooling all of these 141 Z-scored values into a single distribution. A mean and standard deviation were then computed for the pooled scores, and these (the $\overline{X}$ and S) were then used to develop the expected frequencies in each of nine cells as below:

Cell Dimensions (in standard deviation)	Expected Frequency	Observed Frequency
-2.00 to -1.75	3.059	1
-1.75 to -1.50	11.98	15
-1.50 to -1.00	22.24	30
-0.50 to 0.50	58.88	51
0.50 to 1.00	21.6	19
1.00 to 1.50	12.32	15
1.50 to 1.75	3.25	6
1.75 to 2.00	1.847	4
($\overline{X}$ = 0.00612, S = 0.909337)		

A chi square test yields a value of 10.574, meaning that the data does not fit a normal distribution. However, it should be noted that the actual shape of the curve is that of a normal distribution - but a very leptokurtic one. A number of transformations may serve to normalize this data. Analysis of like data from other test systems yielded similar results. It is likely that the censored nature of these measurements (*i.e.*, no values beyond 30 minutes are recorded) is the major cause of the data distribution being as it is.

As a result, the following steps should be taken with time to incapacitation and time to death data:

1. Median and range will be used as descriptive statistics. All graphical representations will also include individual data points to better represent dispersion.
2. Hypothesis testing will be performed by a multiple sum of ranks method.
3. A number of transformations will be explored to see if they normalize their data.

4. As additional data is generated, it will be added to the statistical analysis pool.

Similar analyses were perfrmed on per cent mortality and per cent incapacitation data from both the CMIR and other test systems along with a probit transformation linear fit test (Gad and Weil, 1988). Both of these sets of data were not significantly different from normal at the 0.05 level, but showed significant deviations from linearity when it was attempted to correlate them with exposure concentration for several agents.

Finally, the manner in which the biological effects of thermolysis products are now analyzed and are proposed to be used needs to be rethought. As was pointed out early in this volume, the results of the currently employed test methods focus on lethality (particularly median lethality, or the LC_{50}) as the endpoint of concern. The minimum lethal concentrations (not the median) is probably a more appropriate criterion for concern over lethality - particularly if the tests are being used as a screen. Respiratory and cardiovascular impairment, as well as incapacitation, at least should be of equivalent concern. Additionally, as was pointed out in the preceding paragraphs, because of the way the tests are performed and the measurements taken, these data are neither parametric nor normal in nature. Accordingly, a classification or ranking scheme is more suitable for the kinds of data currently available.

But if a ranking scheme such as the one proposed is used, there is a good correlation of results from test system to test system. if one or more of these tests is to be used in regulation of materials, then a well though out set of point standards need to be formulated. These would be critical values on which there is wide agreement that any material "worse" than a set level would not be suitable for certain uses. Such critical values should not be the LC_{50} of wood (for example), but rather a value which represents a truly significant increase in hazard (such as a tenth that of wood for materials to be used in large volumes in furnishing, paneling and such).

REFERENCES

Cornish, H. H., Hahn, K. J. and Barth M. L (1975) Experimental Toxicology of Pyrolysis and Combustion Hazards. Environ. Hlth. Perspect. **11**: 191-196.

Gad, S. C. (1988) An Approach to the Design and Analysis of Screening Studies in Toxicology. *Journal of the American College of Toxicology* **7**: 127-138.

Gad, S. C. and Smith, A. C. (1984) Influence of Heating Rates on the Toxicity of Evolved Combustion Products: Results and a System for Research. *J. Fire Sciences* **1**: 465-479.

Gad, S. C. and Weil, C. S. (1988) *Statistics and Experimental Design for Toxicologists.* Telford Press, Caldwell, NJ, 2nd Edition.

Green, R. J. S., Hume, V. and Kumar, S. (1976) Burning Plastics: Mechanisms of Smoke Formation. *Fire and Mater.* **1**: 36-41.

Hilado, C. J. (1970) The Effect of Chemical and Physical Factors on Smoke Evolution from Polymers. *J. Fire Flam.* **1**: 217-238.

Hilado, C. J. and Brauer, D. P. (1979) Toxicity of Pyrolysis Gases From Thermal insulation Materials. *J. Therm. Insul.* **2**: 90-101.

Hilado, C. J., Casey, C. J. and Schneider, J. E. (1979) Effect of Pyrolysis Temperature on Relative Toxicity of Some Plastics. *Fire Technol.* **15**: 122-9.

Zapp, J. A. (1951) *The Toxicity of Fire.* Chemical Corps., Army Chemical Center, Maryland; Medical Division Special Report No. 4.

Zikria, B. A., Ferrer, J. M. and Floch, H. F. (1972) The Chemical Factors Contributing to Pulmonary Damage in "Smoke Poisoning". *Surgery* **71**: 704-409.

Index

A

acetic acid, 51
acrolein, 5, 57, 64, 74, 91-92, 150, 155, 169
acrylic, 78, 153, 174
acrylonitrile, 153, 171, 176, 199
alumina hydrates, 190
aluminum compounds, 185
alveoli, 36-39, 55, 61
Ames test, 12
ammonia, 51, 57, 64, 160
anoxia, 53, 59
antimony compounds, 181
antimony trioxide, 190
antistatic agents, 180
apnea, 114
auto ignition temperature, 83, 122

B

behavioral methods, 90-92
boron compounds, 184-185
Boyle's law, 25
bromine compounds, 183
bronchioles, 36, 38, 55
building code, 18, 21, 97, 127, 129-131, 134, 140, 194

C

carbon dioxide, 29, 32-35, 39-40, 42-43, 58, 60, 64, 74, 80, 120, 127, 160
carbon monoxide, 2-3, 5-6, 10, 14, 16, 53, 56, 62, 64-69, 71, 79-80, 120-121, 124, 126-127, 150, 153, 155, 160, 162, 165, 196
carboxyhemoglobin, 2-3, 15, 56, 60, 66, 161
cellulosics, 155, 164, 167, 184
Charles' law, 25
chemoreceptors, 39
chlorine, 57, 78, 157, 182-183, 187-188
chlorine compounds, 182
colorants, 179, 187
combustion atmospheres, 21, 67-68, 70, 72-74, 80-81, 86, 88, 91, 93-98, 100-102, 120, 158, 161-163, 197
combustion products, 9, 11, 14-15, 20-21, 57-58, 63, 79-80, 82, 85-87, 96-97, 100, 103, 113, 119-120, 124-128, 140, 151, 153, 155, 164, 172-175, 177, 189-191, 201
core temperature, 49
coupling agents, 178
curing agents, 180
cytochrome system, 3, 29

D

Dalton's law, 35
decomposition temperature, 5, 85, 112, 114, 153, 167, 169
dose response, 82
Douglas fir, 63, 88, 94, 105, 114, 148, 150, 162-163

E

ECT50, 59
erythrocyte, 42-44
exercise wheel, 74, 91
expanded polystyrene, 20
exposure chamber, 20, 84-88, 91, 100, 102, 104, 106, 109, 112, 119, 121-122, 195

F

fatality rates, 1
Federal Trade Commission, 17
fillers, 104, 178, 185
fires, 1-3, 6, 13-17, 19, 21, 56-58, 63-64, 66-67, 77, 79-82, 85-86, 107, 121, 125, 129, 147, 176, 195-197
flame retardants, 2, 77, 171, 175, 177-178, 181-184, 186-192, 198-199
flaming, 12, 15, 63, 83, 100-101, 108, 118, 132, 149, 153, 157-158, 161-162, 171, 175, 183, 188, 190, 197, 199
flashover, 18
fluoropolymers, 156-159, 173
furnace, 84-88, 100-101, 103-104, 106, 109, 111-112, 116-117, 119-122, 132-133, 156, 174

G

gas exchange, 36, 39, 60
Gas Laws, 25
glottis, 36

H

heart rate and cardiac output, 49
hemoglobin, 2-3, 15, 40-44, 56, 60, 66, 121, 161
histopathological changes, 94, 157
hydrogen chloride, 51, 57, 67, 73, 79, 125-126, 173
hydrogen cyanide, 3, 10, 14, 64, 66, 70-71, 79-80, 126, 150, 166, 174, 176
hydrogen fluoride, 51, 157
hydrogen sulfide, 51
hypoxia, 3, 56, 59-60, 64, 92, 98

I

ignitability, 21, 195
inhalation toxicology, 23, 25, 27, 29, 31, 33, 35, 37, 39, 41, 43, 45-47, 49, 51, 53, 55, 57, 59-62, 79, 87, 125, 173
isocyanates, 3, 5, 64, 98, 102, 167

L

larynx, 36, 61, 93
latex, 171, 176, 185
LC_{50}, 9-10, 60, 70-71, 74, 82, 87-88, 94, 96-97, 101, 114-115, 117, 120, 122, 135, 137-138, 148-149, 153, 155, 157-158, 161-165, 171-172, 190, 193, 195, 198, 201
LCT_{50}, 60, 159
LT_{50}, 60, 88, 116, 122, 148, 153, 155, 163
lung, 11, 14, 24, 29, 35-40, 42, 45, 50-55, 57-62, 78, 107, 157, 167, 169, 173, 176, 198
lung volume, 39

M

mathematical modeling, 3, 6, 13-14, 79, 126, 197
mutagenic compounds, 12
myoglobin, 41-42

N

narcosis, 61
natural products, 2
nitrogen dioxide, 51, 57
nitrogen oxides, 57, 64, 198

O

ozone, 51, 57

P

pharynx, 29, 36, 61
phenolics, 163, 167, 174, 185
phosgene, 51, 57, 167, 169, 173
phosphorus compounds, 184
plasticizers, 177, 179-180
pneumonoconiosis, 55
polyacrylonitrile, 171, 176
polyamides, 159, 163, 191
polycarbonates, 164, 184
polychloroprene, 171, 182
polyethylenes, 164-165
polyimides , 163
polymer fume fever, 11, 16, 158
polyphenylene oxide, 172, 183-184
polyphosphazene, 172, 175
polypropylenes, 164-165
polystyrenes, 165
polytetrafluorethylene, 11
polyvinyl chloride, 11, 63, 79, 125, 169, 174, 176, 179
polyvinylchloride, 3, 125, 173
postmortem studies, 64
processing aids, 180
pulmonary irritation, 58, 112
pyrolysis, 1, 14-16, 63, 77, 80, 83, 104, 120, 126-128, 153, 164, 172-176, 181, 190-191, 199, 201-202

R

regulatory approaches, 129
regulatory control, 138
respiratory control, 40
respiratory system, 32, 36, 40, 50, 62

S

sensory irritation, 11, 13, 58, 61, 72, 79, 87, 93-94, 100, 102, 112-113, 115-116, 124-125, 127, 172
silicone polymers, 166, 174
smoke, 1, 3, 9, 13-21, 53, 56-58, 63-64, 66-67, 72, 77-78, 80, 85-88, 90-91, 93-94, 96, 99-100, 108-109, 113, 117, 119, 122, 124-127, 129-133, 135-137, 139, 141, 143, 145, 155, 157, 159-160, 167, 171-172, 174-175, 185-186, 190-196, 198-199, 201-202
smoke density, 77-78, 129-130, 198
smouldering, 83, 197
sprinklers, 129
stabilizers, 179
stain resistant agents, 180
stress index, 94, 113-114
stress response, 49
structure activity relationship, 6
sulfur dioxide, 57, 61, 64, 167
synthetic polymers, 1-3, 13, 64, 126, 151, 174-176, 191

T

Tefzel 200, 95, 157-158
test atmosphere, 5, 24, 62, 131
thermal decomposition, 1, 5, 11, 13, 15-16, 63-64, 79, 83, 85, 96, 106, 115-117, 119-120, 124-125, 127, 153, 170, 172-177, 189-190
thermolysis, 12, 83-84, 102, 150, 158-160, 164, 173, 177-178, 197-198, 200-201

thoracic cavity, 36, 38
toxicity measurements, 86-87, 101
trachea, 24, 36, 53, 55, 61, 66, 93

U

urethane foam, 15, 18-19, 21, 126, 173-178, 183-186, 189, 191-192
urethanes, 166-168, 171, 174, 189

V

vinyls, 169, 178

W

warning devices, 129

X

xenobiotic chemical, 9